Current Chinese Economic Report Series

The Current Chinese Economic Reports series provides insights into the economic development of one of the largest and fastest growing economies in the world; though widely discussed internationally, many facets of its current development remain unknown to the English speaking world. All reports contain new data, which was previously unknown or unavailable outside of China. The series covers regional development, industry reports, as well as special topics like environmental or demographical issues.

More information about this series at http://www.springer.com/series/11028

Center for Macroeconomic Research at
Xiamen University

China's Macroeconomic Outlook

Quarterly Forecast and Analysis Report,
February 2018

 Springer

Center for Macroeconomic Research at
   Xiamen University
Xiamen University
Xiamen, Fujian, China

ISSN 2194-7937   ISSN 2194-7945   (electronic)
Current Chinese Economic Report Series
ISBN 978-981-13-1004-1   ISBN 978-981-13-1005-8   (eBook)
https://doi.org/10.1007/978-981-13-1005-8

Library of Congress Control Number: 2018945447

Printed on acid-free paper

This Springer imprint is published by the registered company Springer Nature Singapore Pte Ltd.
The registered company address is: 152 Beach Road, #21-01/04 Gateway East, Singapore 189721,
Singapore

Preface

This report is a partial result of the China's Quarterly Macroeconomic Model (CQMM), a project developed and maintained by the Center for Macroeconomic Research (CMR) at Xiamen University. The CMR, one of the Key Research Institutes of Humanities and Social Sciences sponsored by the Ministry of Education of China, has been focusing on China's economic forecast and macroeconomic policy analysis, and it started to develop the CQMM for the purpose of short-term forecasting, policy analysis, and simulation in 2005.

Based on the CQMM, the CMR and its partners hold press conferences to release forecasts for China' major macroeconomic variables. Since July 2006, 23 quarterly reports on China's macroeconomic outlook have been presented and 12 annual reports have been published.

This report, the twenty-fourth quarterly report, has been presented at the Forum on China's Macroeconomic Outlook and Press Conference of CQMM on February 27, 2018. This conference was jointly held at Beijing, China by Center for Macroeconomic Research at Xiamen University, and Economic Information Daily at Xinhua News Agency.

Xiamen, China   Center for Macroeconomic Research at Xiamen University

Acknowledgements

According to the Chinese Pinyin order of their names, the 128 experts who joined this questionnaire survey were: Bai Peiwen, Chang Xin, Chen Changbing, Chen Gong, Chen Heng, Chen Jianbao, Chen Jinmei, Chen Kunting, Chen Langnan, Chen Menggen, Chen Shoudong, Chen Xikang, Chen Yanbin, Chen Zhiyong, Dai Kuizao, Dong Ximiao, Fan Conglai, Fan Ziying, Gao Bo, Geng Qiang, Guo Qiyou, Guo Xibao, Guo Xiaohe, Guo Zhiyi, Han Zhaozhou, Hu Ridong, Hwa Erh-cheng, Huang Jianfeng, Huang Maoxing, Huang Xianfeng, Kan Kamhon, Jian Xinhua, Jiang Yongmu, Jin Tao, Li Chunqi, Li Haizheng, Li Jianwei, Li Jun, Li Shi, Li Xuesong, Li Yingdong, Leung Ka Yui Charles, Lin Xuegui, Liu Jianping, Liu Jinquan, Liu Qiongzhi, Liu Shiguo, Liu Yunzhong, Liu Zhibiao, Lui Hon-kwong, Ma Ying, Peng Suling, Qi Jindong, Qin Wei, Chu Wan-wen, Ren Baoping, Ren Ruoen, Shao Yihang, Shen Guobing, Shen Kunrong, Shen Lisheng, Shi Kang, Shi Junyi, Shi Jinchuan, Su Jian, Sun Wei, Tang Jijun, Wang Changyun, Wang Tongsan, Wang Yida, Wang Cheng, Wang Dashu, Wang Guocheng, Wang Jiping, Wang Junbin, Wang Junbo, Wang Liyong, Wang Susheng, Wang Xi, Wang Yanhang, Wang Yuesheng, Wen Chuanhao, Wu Kaichao, Wu Xinru, Wu Kangping, Xiao Xingzhi, Xie Danyang, Xie Di, Xie Pan, Xu Xianxiang, Xu Yifan, Xu Bin, Xu Wenbin, Yan Ping, Yang Chengyu, Yang Cuihong, Yang Zhiyong, Yin Heng, Yi Xianrong, Yin Xiongmin, Yu Li, Yu Zuo, Yuan Fuhua, Zang Xuheng, Zeng Wuyi, Zhang Donghui, Zhang Fan, Zhang Liqun, Zhang Liancheng, Zhang Long, Zhang Mingzhi, Zhang Ping, Zhang Shuguang, Chang Gene, Zhang Yifan, Zhao Minghao, Zhao Xijun, Zhao Xiaolei, Zhao Zhenquan, Zhao Zhijun, Zheng Chaoyu, Zhong Chunping, Zhou Liqun, Zhou Zejiong, Zhu, Baohua, Zhu Jianping, Zhu Qigui, and Zhuang Zongming.

The experts who joined this questionnaire survey are from institutions like Baoshang Bank, Ministry of Finance, National Bureau of Statistics, Development Research Center of the State Council, Research Institute of Hengfeng Bank, Research Institute of Minsheng Bank, Tianze Institute of Economic Research, International Department, Central Committee of CPC, Academy of Mathematics and Systems Science of CAS, Center for Forecasting Science of CAS, National

Academy of Economic Strategy of CASS, Institute of Finance and Banking of CASS, Institute of Economics of CASS, Institute of World Economics and Politics of CASS, Institute of Quantitative and Technical Economics of CASS, China Banking Association, Taiwan Chung-Hua Institution for Economic Research, Taiwan Academia Sinica and universities like Anhui University of Finance and Economics, Peking University, Beihang University, Beijing Normal University, Chongqing Technology and Business University, Dongbei University of Finance and Economics, Fujian Normal University, Fudan University, Guangxi University, Hunan University, East China Normal University, Huaqiao University, Huazhong University of Science and Technology, Jilin University, Jinan University, Lanzhou University, Liaoning University, Georgia Institute of Technology, Nanjing University, Nankai University, Qingdao University, Tsinghua University, Shandong University, Shaanxi Normal University, Shanghai University of Finance and Economics, Shanghai University of International Business and Economics, Shanghai Jiao Tong University, Capital University of Economics and Business, Sichuan University, Taiwan University, Tianjin University of Finance and Economics, Tianjin University of Commerce, Wuhan University, Xi'an Jiao Tong University, Northwest University, Southwestern University of Finance and Economics, Xiamen University, City University of Hongkong, University of Hongkong, Hongkong University of Science and Technology, Lingnan University in Hongkong, Chinese University of Hongkong, Yunnan University of Finance and Economics, Zhejiang University of Finance and Economics, Zhejiang University, Zhejiang University of Technology, Renmin University of China, Zhongnan University of Economics and Law, Central South University, China Europe International Business School, Sun Yat-sen University, Central University of Finance and Economics, etc.

For the active participation and insights of all abovementioned experts, we wish to extend our deep thanks.

Contents

Contributors

Principal Investigator

Gong Min

Research Team Members

Li Wenpu, Lin Zhiyuan, Liu Yu, Lu Shengrong, Chen Guifu, Yu Changlin, Wang Yanwu, Li Jing, Wu Huakun

The Chinese edition of this report is contributed by Gong Min, Wang Yanwu, Wu Huakun, Lu Shengrong, Li Wenpu, and Yu Changlin. It is translated by Liu Yu, Yu Changlin, and Zhong Yu. The raw data are processed by Wu Huakun.

Executive Summary

In 2017, the steady growth of economic growth of China continued to be consolidated. The gross domestic product (GDP) grew by 6.9%, an increase of 0.2 percentage points over the previous year, which were attributed to the rapid growth of infrastructure investment and the rapid rebound in the growth of imports and exports. On the one hand, the economic structure continued to be optimized, the share of tertiary industry increased further, and the share of high-end manufacturing investment and output continued to rise. On the other hand, the leverage ratio of financial and nonfinancial sectors rose rapidly and stayed at high levels, increasing the systemic risk of economy. This not only indicates that the factors promoting economic growth were more complex but also indicates that the prospects for economic growth were more uncertain.

While economic growth of China had stabilized, it had become increasingly evident that some factors might have weakened the growth since 2015: First, the contribution rate of industrial sector to GDP growth had dropped significantly. This was largely a reflection of the low efficiency of industrial production and the need for further advancement in industrial restructuring. Second, the growth of investment in fixed assets continued to fall, and the investment structure was further out of balance. The share of investment in state-owned enterprises continued to rise, while the share of private investment continued to fall; the share of manufacturing investment continued to fall, while the share of infrastructure investment continued to rise and the share of real estate investment remained high, indicating that the investment structure had not been adjusted in the direction of improving investment efficiency. Third, the stabilization of industrial production benefited from the expansion of state-owned and state-controlled enterprises, though the increase in investment and profits of private enterprises in the competitive sector was relatively mild. Although the different trends of investment and profit growth between state-owned and private enterprises had ensured the growth of tax revenue in the short term, it had been detrimental to the improvement of industrial production efficiency and the upgrading of industrial structure in the long term. Worstly, it would further reduce the allocation efficiency of resources among industries. Fourth, in the past 2 years, although the new loans had been expanded year by year, the real economy accounted for less than 50% of the new loans. Financial resources were excessively concentrated in the real estate industry, infrastructure, and the state-owned enterprises. With the existence of various investment barriers, the financial needs of the real economy, especially private enterprise investment, could not be fully and effectively guaranteed.

In 2018, reducing leverage ratio of the financial and nonfinancial sectors and preventing major financial risks are still the core concerns of various macroeconomic policies. First of all, the regulatory oversight of local finances will continue to be strengthened, the loopholes through which local governments could borrow money abnormally had basically been blocked and the high-speed expansion of infrastructure investment will be curbed; Second, the regulatory authorities begin to crack down on all kinds of financial chaos, especially the various types of irregularities on real estate development loans will be severely punished, which will largely inhibit the growth of real estate investment; Finally, combined with the reform of state-owned enterprises with regard to policies such as reducing production overcapacity and production cost, the process of reducing leverage ratio of state-owned enterprises will also be substantially advanced. These measures to prevent and control financial risks will inevitably aggravate the downward pressure on investment growth. If private investment growth cannot rebound substantially, then the deceleration of investment in fixed assets will inhibit economic growth. With the economic slowdown, the demand for infrastructure expansion, and the reliance of local government on the revenue from land sales, the financial sector's preference for state-owned enterprises and the real estate industry may reappear. Consequently, it is difficult to reverse the situation in which a large amount of capital is idling within the financial system without entering the real economy, and the risk of financial and nonfinancial state-owned enterprises is more difficult to be eliminated. Therefore, while preventing and controlling financial risks, it is necessary to further investigate how China can effectively stimulate the growth of private investment, improve the efficiency of resource allocation, and thus promote a series of issues such as structural adjustment, transformation and upgrading of manufacturing industry, and improving the growth of labor productivity.

Forecasts based on the CQMM model show that the sustained growth of the global economy will continue to stimulate the foreign trade of China in the next 2 years, but the slowdown in domestic investment growth will continue to inhibit economic growth of China. On the whole, China's economy is expected to maintain a level of 6.50% or more in the next 2 years, and the trend of "stable and slowing down slightly" in economic growth will continue.

(a) It is expected that real GDP of China will increase by 6.73% in 2018, a decrease of 0.17 percentage points over the previous year; CPI will reach 2.13%, an increase of 0.53 percentage points over the previous year; PPI will increase to 4.64%, a decrease of 1.66 percentage points over the previous year. The price level remains in a controllable range, there is no obvious inflation risk, and the gap between the PPI and CPI will also be reduced.

(b) The strengthening of financial market supervision and control over financial system risks, as well as the control the debt risks of local governments, coupled with the slow rebound of private investment, will continue to curb investment growth in the next 2 years. Nominal fixed asset investment (excluding rural households) is expected to increase by 6.57% in 2018, a decrease of 0.63 percentage points over the previous year.

(c) The growth rate of disposable income of urban and rural residents may decline slightly, resulting in a decrease in the growth of household consumption. In 2018, the actual growth rate of resident consumption will be about 6.79%, a decrease of 0.44 percentage points over the previous year. The total retail sales of nominal social consumer goods are expected to increase by 10.60% in 2018, an increase of 0.30 percentage points over the previous year.

(d) The recovery of the world economy will continue to drive the growth of foreign trade of China in 2018. The total value of export (current US dollar value) is expected to increase by 9.65%, an increase of 1.75 percentage points over the previous year; and the total value of imports (current US dollar value) will increase by 12.32%, a decrease of 3.58 percentage points over the previous year. Stabilization of the RMB against the US dollar will continue to ease pressure on China's capital outflow. It is expected that the scale of foreign exchange reserves will expand to US$3.33 trillion in 2018.

In recent years, China's economic, financial, and nonfinancial sectors have rapidly increased their leverage ratios, leading to an increase in systemic risks in the economy. Among them, due to the existence of "soft budget constraints" and "implicit guarantees" from local government departments, the leverage ratio of state-owned enterprises was ease to rise but hard to fall, creating a ratchet effect. Although, at the end of 2015, the Central Economic Work Conference had already proposed to reduce leverage ratio and to reduce production overcapacity, etc., the deleveraging process was not optimistic: First, the leverage ratio of nonfinancial enterprises had declined recently, but it was still at a relatively high level. Second, in nonfinancial state-owned enterprises, the leverage ratio of state-owned industrial enterprises had declined, but the leverage ratio of all state-owned enterprises was still at a high level. Finally, the progress of deleveraging in the financial sector was still not optimistic. In the past years, the central bank strictly controlled the growth of money supply, tightened the supply of funds, and reduced the financial leverage ratio; on the other hand, it strengthened financial supervision and strictly controlled the rapid expansion of internal financial derivatives in the financial sector, especially the financial expansion of banks and nonbanking institutions. Consequently, it was difficult to reverse the situation in which a large amount of capital was idling within the financial system without entering the real economy. However, although the new loans in the past 2 years had been expanding gradually, the proportion of loans to the real economy is still less than 50%.

To prevent systemic financial risks, it is necessary to effectively control the macro-leverage ratio, but this will inevitably result in a decline in the growth of state-owned investment and shake the important pillars of the current growth, especially in the context of slow rebound of private investment growth. In view of this, the research team used the CQMM model to simulate the possible macroeconomic impact of the deleveraging policy of state-owned enterprises in the current 2 years. Focus on simulating the impact of deleveraging of state-owned enterprises on investment growth and economic growth, and provide a quantitative analysis basis for the macro-effect with policy suggestions.

The research team designed two types of simulation scenarios:

Scenario 1: Assuming a substantial advancement of the deleveraging policy in the next 2 years, the leverage ratio (asset–liability ratio) of state-owned industrial enterprises will fall to 56% within eight quarters, basically returning to the level before the outbreak of the financial crisis in 2007. The setting of this simulation scenario can quantify the extent to which the state-owned enterprises' deleveraging policy may have a negative impact on investment growth and economic growth.

Scenario 2: On the basis of Scenario 1, to ensure that the economic growth rate is stable at a level of about 6.5%, then how much should private investment rebound? The setting of this simulation scenario can quantify the "compensation effect" of private investment on the deleveraging policy and provide support for empirical analysis for policy authorities to formulate relevant policies to promote the recovery of private investment growth.

The simulation results show that when other conditions remain unchanged, the reduction in leverage ratio of state-owned enterprises will drastically reduce the investment growth rate of state-owned enterprises, which in turn will reduce the growth rate of fixed asset investment in the entire society and further reduce the growth rate of GDP of China. However, if the economic growth rate in the next 2 years is to be stabilized at a level of 6.5%, then private investment growth will rebound to 13.35% and 14.42%, respectively, in the next 2 years, and the expansion of private investment will effectively compensate for the shrinking of the state-owned investment. The state-owned investment, which has been declining due to reducing leverage ratio, has achieved steady economic growth. Therefore, the research team proposed that while promoting the deleveraging of state-owned enterprises, it must be able to effectively stimulate the growth of private investment. Otherwise, it will be difficult for both leverage reduction and steady growth to take effects.

To some extent, the growth of nongovernment investment has been difficult to rapidly rebound since 2015, which is the main reason why macroeconomic problems have not been fundamentally resolved. First of all, it is difficult for the private investment to quickly rebound, and local governments at all levels have to increase investment in infrastructure; at the same time, the financial sector also has to expand financial supply to state-owned enterprises. This not only increased the leverage ratio of nonfinancial state-owned enterprises but also further distorted the investment structure. In addition, the long-term policy in favor of state-owned enterprises had inevitably "squeezed out" the financial resources available for private investment, further inhibiting the rebound of private investment.

Second, the decline in private investment growth is directly reflected by the shrinking growth in manufacturing investment. This not only inhibits the expansion of industrial production but also is not conducive to the upgrading of industrial structure and industrial efficiency. Further, it will curb the growth of profits of industrial enterprises and slow down the growth of fiscal tax revenues. Consequently, the governments' reliance on revenue from land sales is growing stronger, pushing up the bubble in the real estate market.

Third, it is difficult for nongovernment investment to rebound rapidly, which not only inhibits the growth of investment in fixed assets but also leads to further imbalances in investment structure and further reduces investment efficiency. This imbalanced investment structure is undermining the economic efficiency of investment and increasing the fragility of the financial system.

Finally, in the near future, the rapid growth of labor productivity, especially in the manufacturing industry, will inevitably be an important guarantee for the improvement of labor productivity and the rapid increase in real income of residents. To reverse the declining trend in labor productivity growth, China must rely on the rapid increase in labor productivity in the manufacturing industry. Effective investment expansion (especially private investment) is the necessary means to promote the transformation and upgrading of the manufacturing industry and thus improve labor productivity. Therefore, the current risk of continued decline in the growth rate of private investment is significant, and the decline in the growth rate of investment in manufacturing should be given great attention.

According to the forecast and the policy simulation analysis made by the research team in the next 2 years, the research team believes that the rapid rebound of private investment growth has many favorable conditions: First, the continued growth of the world economy in the next 2 years will inevitably stimulate the export of private enterprises of China, and the export expansion will in turn motivate the growth of its investment; Second, the financial resources released by the state-owned enterprises due to deleveraging and the adjustment of the structure of financial resources to enable them to serve the real economy will largely meet the funding needs of private investment expansion; Third, further deepening reforms will reduce the long-standing investment barriers to private enterprises, which will expand the investment area of private enterprises; Finally, the renovation, cultivation, and improvement of the business environment begun this year will also expand private investment. Based on this, policy authorities should make full use of the opportunity of the domestic and foreign economies in the next 2 years and comprehensively apply various policies and measures. While preventing and controlling financial risks, the policy should focus on stimulating the private economy and promoting the rapid growth of private investment.

Therefore, it is extremely necessary to rectify the financial governance in the next 3 years, and China should also pay attention to adopting appropriate measures to protect the loan demand for investment in manufacturing, with the continuous improvement of the world economy in the 2 years. The intensive expansion and its transformation and upgrading will consolidate the industrial base of economic growth and will accelerate the improvement of labor productivity, and ultimately achieve a long-term rapid increase in the real income of residents. In the long run, only by relying on the market forces and improving the efficiency of the allocation of production factors between different ownership enterprises and different industries can China ensure a rapid increase in labor productivity. While preventing and controlling financial risks, China is possible to stabilize economic growth and achieve development with high quality.

In particular:

(a) Stable growth of economy basically requires the rapid growth of effective investment and the improvement of investment efficiency. While the new economic growth momentum has not yet been found, the pressure of continued declining investment growth should be highly valued. Therefore, while preventing and controlling financial risks, China must start with adjusting the structure of financial resources so that it can fully meet the needs of the real economy, especially the expansion of private investment demand, so that finance can truly serve the real economy.

(b) The rapid improvement in labor productivity requires the intensive expansion of the manufacturing industry and its transformation and upgrading. Its driving force will certainly come from private investment. The current over-concentration of financial resources in the real estate industry and the state-owned enterprises make it difficult for private manufacturing enterprises to obtain financial support. The distortion of the financial structure has resulted in low allocation efficiency. Therefore, while preventing and controlling financial risks, it is necessary to speed up the market-oriented reform of the capital market and improve the efficiency of allocating financial resources among industries so as to promote the adjustment of investment structure.

(c) Monetary policy should maintain "stable and tightening," strictly control the money supply, and actively coordinate macro-prudential supervision and financial stability policies so as to achieve effective prevention and control of financial risks. China will continue to increase the supervision of shadow banking, combine the assessment of macro-prudential assessment system (MPA) to strictly control the expansion of shadow banking, reduce the unregulated financing behavior, and actively guide shadow banking to give full play to the function of serving the real economy.

(d) While reducing the debt ratio of nonfinancial state-owned enterprises, China should speed up the establishment of a modern fiscal and taxation system, form a hard constraint on the budget of local governments (state-owned enterprises), and eliminate the possibility of systemic financial risks. To resolve the soft budget constraints of local governments and to control their ability to expand debt, China must establish compatible mechanism between the fiscal revenue and expenditure of a local government, and establish a standardized bond issuance system.

(e) China must change the old development mode which overstress on maintaining a high level of economic growth rate. The governments should gradually switch their functions and focus on public service and public-government governance so that the market can do market affairs and the governments do governments' affairs.

(f) By improving laws and regulations, and ensuring full protection of state-owned assets, China will privatize some state-owned enterprises in the competitive sector by auction. At the same time, China will steadily promote the reform of the mixed ownership system of state-owned enterprises to promote the competitive field of state-owned monopoly industries, to stimulate private investment.

(g) To stimulate the private investment, it is necessary to boost private entrepreneurs' confidence in the domestic economy by deepening comprehensive reforms, improving the domestic economic environment and the business environment, protecting entrepreneurial spirit and corporate property rights, and supporting entrepreneurs to concentrate on innovation and entrepreneurship.[1]

In summary, the research team believes that effective measures to promote the rapid recovery of nongovernment investment will help solve some of the problems and help advance the structural reforms on the supply side. In the next stage, "promoting reforms through openness, promoting growth through innovation, improving efficiency through competition, and ensuring employment through demand" should be the preconditions for China's economic development toward high quality. To this end, supply-side structural reforms should be proactively promoted to the market of production factors, and through the market-oriented reforms, the financial capital market should be continuously improved, and the allocation efficiency of financial resources should be improved, through the reduction of institutional transaction costs, the promotion of private investment.

---

[1] Premier Li Keqiang of the State Council presided over the State Council executive meeting on June 22, 2017 to listen to reports on the implementation of special inspections on private investment policies. The meeting pointed out that the State Council organized a nationwide inspection of the implementation of the policy to promote the healthy development of private investment and found some problems: some regulations and policies were not matched and uncoordinated, and implementation was not in place; It is difficult for private enterprises to enjoy the same treatment as state-owned enterprises with regard to market access, resource allocation and government services, etc.; financing is difficult and the burden of payment is heavy; some officials are misconduct; and some local governments have lost their trust, which have seriously affected the healthy development of private investment.

Chapter 1
Review of China's Macroeconomy in 2017

In 2017, the gross domestic product (GDP) of China grew by 6.9%, an increase of 0.2 percentage points over the previous year (Fig. 1.1). The steady growth of economy continued to be consolidated. While the economic structure continues to be optimized, the continued stabilization of industrial production, the rapid growth of infrastructure investment and the rebound in the growth of imports and exports have become an important basis for steady economic growth in 2017. However, since 2015, some new features of the macro economy have emerged, such as the widened disparity between the growth rate of GDP and of industrial added value, between the growth of rate of profits of state-owned and private-owned enterprises, and between the growth rate of loan and investment. These not only indicate that the factors behind the promotion of economic growth are more complicated, but also that the prospects for economic growth are even more uncertain.

1.1 The Economic Structure Continues to Be Optimized, and the Contribution Rate of the Tertiary Industry to GDP Growth Had Increased Substantially

In 2017, the economic structure of China continued to be optimized and adjusted to ensure a stable recovery of economic growth. The tertiary industry accounted for 51.6% of GDP, a slight increase of 0.07 percentage points over the previous year. Owing to the steady growth of industrial production, the proportion of secondary industry was 40.5%, an increase of 0.58 percentage points over the previous year, and the proportion of secondary industry was basically the stable. The proportion of primary industry continues to decline (Fig. 1.2).

The continuous increase in the share of tertiary industry to GDP had ensured the growth of new employment. At the same time, the contribution rate of the tertiary industry to GDP growth was also rising. On the other hand, the contribution rate of

© Springer Nature Singapore Pte Ltd. 2018
Center for Macroeconomic Research at Xiamen University, *China's Macroeconomic Outlook*, Current Chinese Economic Report Series,
https://doi.org/10.1007/978-981-13-1005-8_1

Fig. 1.1 Changes in GDP Growth. *Source* CEIC

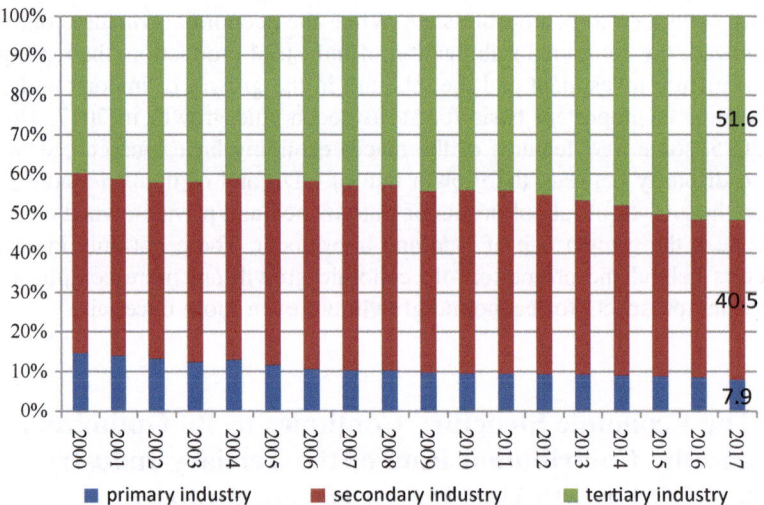

Fig. 1.2 Changes in the share of the three industries in GDP. *Source* CEIC

the secondary industry to GDP growth had continued to decline (Fig. 1.3). Since the beginning of 2015, the contribution rate of the tertiary industry to GDP growth had started to exceed that of secondary industry, and had remained at level of more than 55%; the contribution rate of secondary industry to GDP growth had dropped dramatically to 36.2% in 2017 from 50% averagely before the financial crisis. While the share of secondary industry to GDP is basically stable, its contribution rate to GDP growth had dropped substantially. This fact, to a large extent, reflected the low efficiency of industrial production in China and the need for further advancement of industrial structure adjustment.

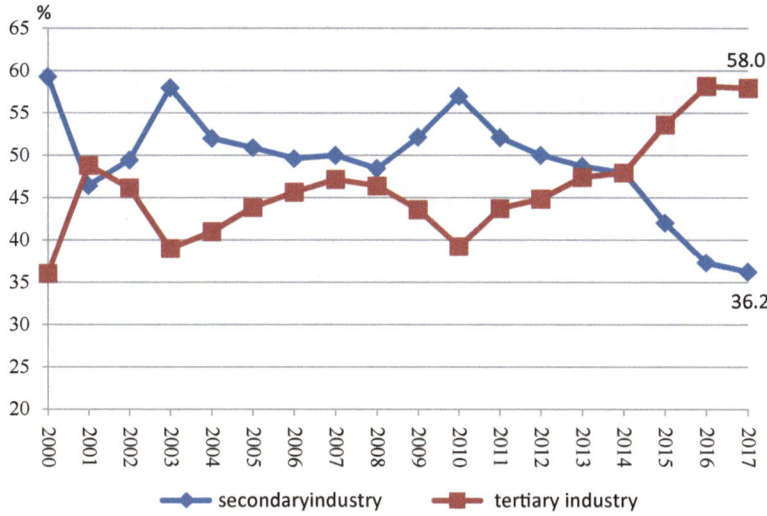

Fig. 1.3 Contribution of two industries to GDP growth. *Source* CEIC

1.2   The Contribution of Foreign Trade to GDP Growth Turned to Positive from Negative

The recovery of the world economy in 2017 effectively boosted the exports and imports of China. At the same time, the "One Belt, One Road" initiative had also strongly driven the growth of foreign trade in the central and western provinces. The contribution of net exports of goods and services to GDP growth reversed from negative to 9.1%, an increase of 15.9 percentage points over the previous year. As investment growth continued to slow down, the contribution of capital formation to GDP growth fell to 32.1%, a substantial decrease of 10.1 percentage points over the previous year. Owing to a relatively stable increase in household income, the final consumption contributed 58.8% of GDP growth, a decrease of 5.8 percentage points over the previous year (Fig. 1.4).

At the same time, final consumption accounted for 53.6% of GDP in 2017, which was the same as that of the previous year. The total capital formation accounted for 44.4% to GDP, a increase of 0.3 percentage points higher over the previous year; net exports of goods and services accounted for 2% of GDP, a decrease of 0.3 percentage points over the previous year (Fig. 1.5). In terms of final demand, household consumption accounted for 39.1% of GDP, a decrease of 0.3 percentage points over the previous year, and government consumption accounted for 14.5% of GDP, an increase of 0.2 percentage points over the previous year. Further, in the consumption of residents, the proportion of urban residents' consumption in GDP had remained at the level of the previous year, which was 30.7%; the proportion of rural residents' consumption in GDP was 8.4%, which was a

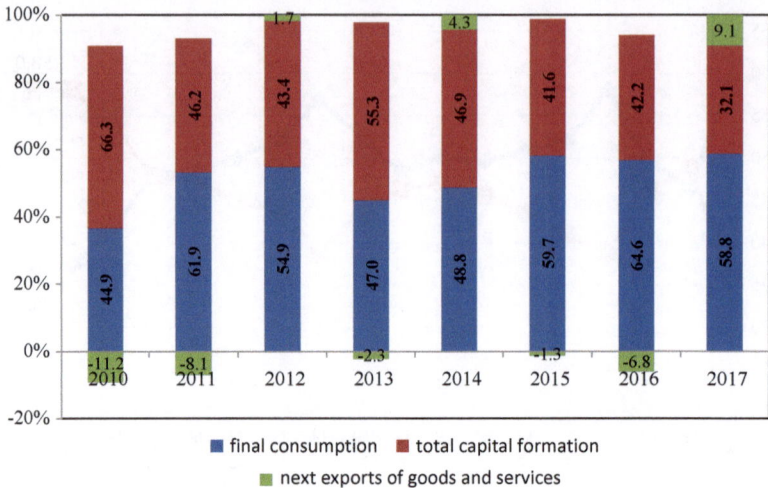

Fig. 1.4 Contribution of GDP growth calculated by expenditure method. *Source* CEIC

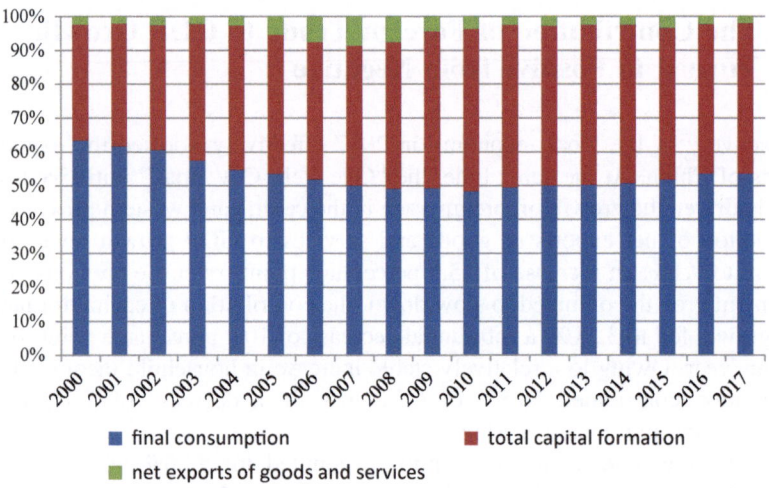

Fig. 1.5 Changes in GDP composition calculated by expenditure method. *Source* CEIC

decrease of 0.2 percentage points over the previous year. It showed that the steady growth of consumption had benefited from the rapid expansion of government consumption to a certain extent; the proportion of rural residents' consumption had declined, to which great attentions must be paid.

1.3   The Growth Rate of Investment in Fixed Assets Continues to Fall, and the Growth Rate of State-Owned and Private Investment Continues to Diverge

In 2017, investment in fixed assets increased by 7.2%, a decrease of 0.9 percentage points over the previous year. Since the beginning of 2010, the investment growth had continued to decline. The disparity between investment growth and GDP growth has also been growing stronger: investment growth had been consistently low in the past three years, but GDP growth had been stable (Fig. 1.6). From a positive perspective, the disparity between investment growth and GDP growth might indicate changes in the investment structure and the dynamic structure underlying the GDP growth. However, the ownership structure of enterprises was reflected in the continuous increase in the share of investment by state-owned enterprises and the continuous decline in the share of private investment; the industrial structure of investment was reflected in the continuous decline in the share of investment of manufacturing in total investment, the rising share of infrastructure investment, and the maintenance of the high level of the share of real estate investment. These characteristics showed that the investment structure had not been adjusted toward the improvement of investment efficiency. Therefore, under the condition that the proportion of investment in GDP had been basically stable, and the GDP of per unit of investment had been declining, the disparity

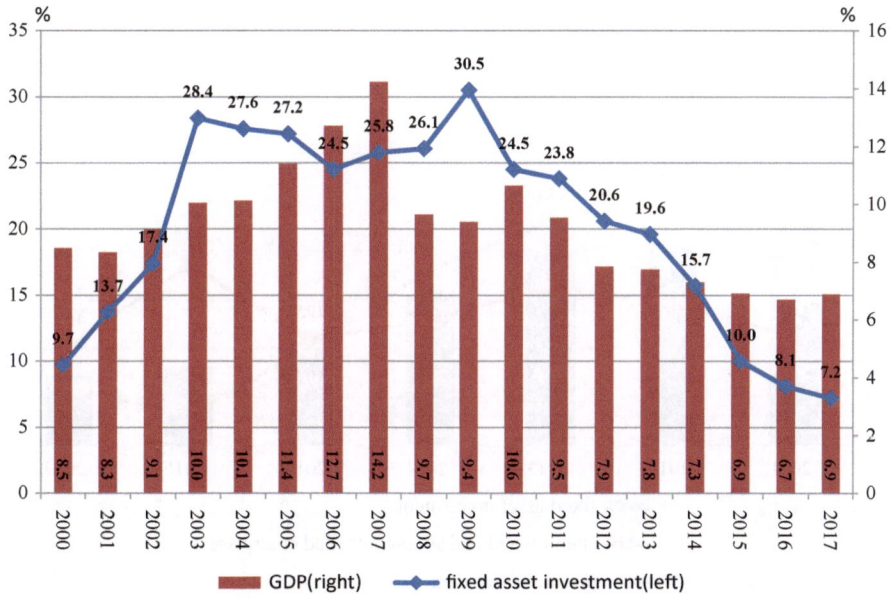

Fig. 1.6 Growth rate of real GDP growth and fixed asset investment growth. *Source* CEIC

between investment growth and GDP growth would be unsustainable, and GDP growth would be deterred by the decline of investment growth.

First of all, from the perspective of the ownership structure of investment (Fig. 1.7), investment in state-owned and state-controlled enterprises increased by 10.1% in 2017, a decline of 8.6 percentage points over the previous year, of which state-owned enterprises' investment grew by 9%, an increase of 15.7 percentage points over the previous year. On the other hand, private investment grew by 6.0%, an increase of 2.8 percentage points over the previous year. The investment growth of Hongkong, Macao and Taiwan investment enterprises had contracted by 4.0%, a drop of 22.5 percentage points over the previous year; the investment growth of foreign-invested enterprises had contracted by 2.7%, a drop of 15.1 percentage points over the previous year. It showed that the investment growth of state-owned enterprises had slowed the decline of investment growth to a certain extent, and it was still difficult for private investment growth to rebound quickly. The disparity in the growth rate of state-owned enterprises and private investment since 2015 had continued. In 2017, the state-owned and state-controlled enterprises' investment accounted for 36.9% of total investment, which was 1.1 percentage points higher than the previous year; private investment accounted for 60.4% of total investment, a decrease of 0.8 percentage points over the previous year and a sharp drop of 3.8 percentage points over 2015. It reached the highest in 2015.

The sluggish growth in private investment had prompted local governments, state-owned and state-controlled enterprises to increase investment to ensure investment growth. In this context, the huge liquidity injected into the economy for

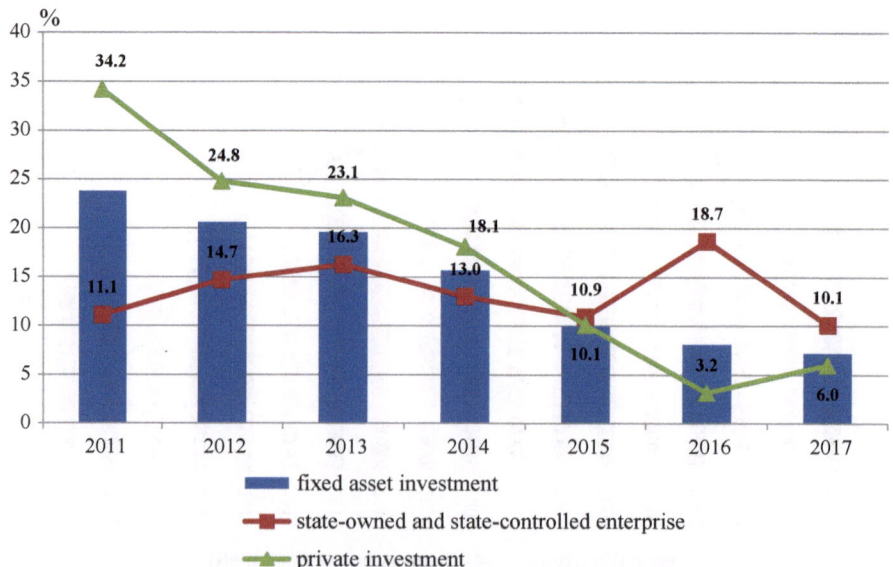

Fig. 1.7 Changes in fixed asset investment growth (classified by ownership). *Source* CEIC

the purpose of stabilizing growth had rapidly increased the high debt of non-financial state-owned enterprises due to the long-term tilt of the financial policy toward state-owned enterprises and the "soft budget constraints" of state-owned enterprises. At the same time, with the "implicit guarantee of the government", the funds released by the loose monetary policy had strongly supported the real estate enterprises, local financing platforms, and the state-owned zombie enterprises (with high-pollution, high-energy-consuming and production over capacity) through the shadow banking system, which in turn pushed up the leverage ratio of the government and state-owned enterprises. As a result, the debt ratio of the non-financial sector and the financial sector had rapidly expanded, which increases the possibility of an outbreak of economic systemic risks.

Secondly, from the perspective of the industry structure of investment, at the industrial level, it is difficult for the private investment to rebound quickly, the growth rate of investment in the secondary industry continues to slowdown, and the share of investment of secondary industries also continues to decline. Investment of primary industries in 2017 increased by 11.8%, an increase of 9.3 percentage points over the previous year, and it accounted for 3.3% of total investment, a slight increase of 0.1 percentage points over the previous year; investment of secondary industries increased by 3.2%, a decline of 0.3 percentage points over the previous year and it accounted for 37.3% of total investment, a decrease of 1.5 percentage points over the previous year; investment in the tertiary industry increased by 9.5%, and the growth rate decreased by 1.4 percentage points over the previous year, and it accounted for 59.4% of total investment, an increase of 1.4 percentage points over the previous year.

At the industry level, the suppression of production overcapacity had increased the prices of upstream resource to a certain extent. Although the investment growth of the mining industry continued to shrink by 10.0% in 2017, the growth rate of the mining industry accelerated by 10.4 percentage points over the previous year, and the proportion of investment of the mining industry to total investment declined to 1.5%, a decrease of 0.3 percentage points over the previous year. Affected by the rapid rebound of non-government investment, manufacturing investment increased by 4.8%, a decrease of 0.6 percentage points over the previous year, and the proportion to total investment continued to drop to 30.7%, a decrease of 0.8 percentage points over the previous year. Under the influence of various real estate management and control policies, real estate investment increased by 3.6%, a decline of 3.2 percentage points over the previous year and it accounted for 22.1% of the total investment, a decrease of 0.6 percentage points over the previous year. Infrastructure investment continued to grow rapidly, which grew by 19.0%, an increase of 1.6 percentage points over the previous year, and it accounted for 22.2% of the total investment, an increase of 2.2 percentage points over the previous year (Fig. 1.8). The share of investment in manufacturing had continued to decline, and the share of investment in infrastructure and in real estate had remained at a high level, which had not only reduced investment efficiency, but also increased the fragility of the financial system.

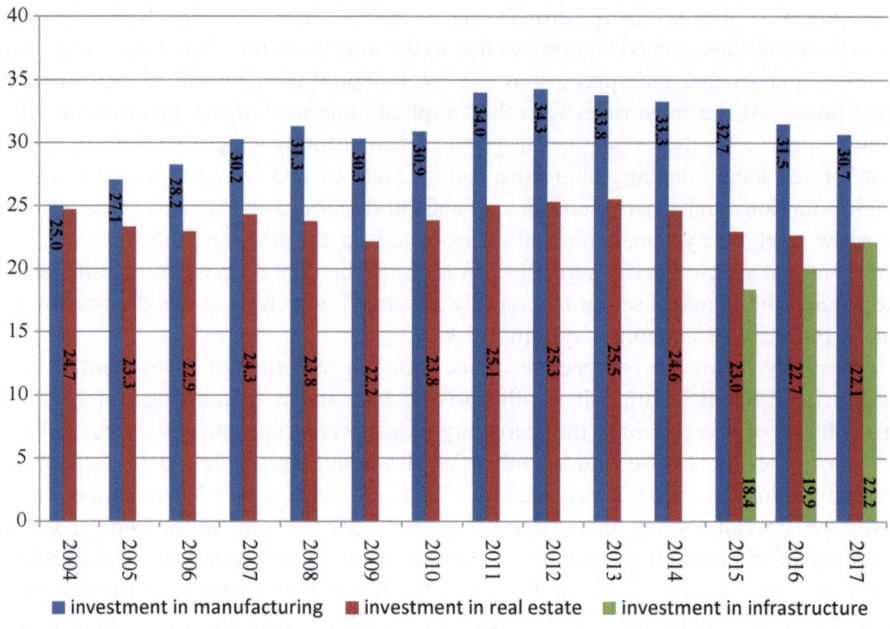

Fig. 1.8 Change in proportion of total investment in major industries. *Source* CEIC

In manufacturing industry, investment in equipment and high-end manufacturing had grown rapidly. Among them, in 2017, investment in general equipment manufacturing increased by 3.9%, investment in manufacturing of special equipment increased by 4.7%, investment in automobile manufacturing increased by 10.2%, investment in railway, shipbuilding, aerospace and other transportation equipment manufacturing increased by 2.9%, investment in computer, communication and other electronic equipment increased by 25.3%. The growth rate was increased by 6.2, 7.3, 5.7, 12.1, and 9.5 percentage points over the previous year respectively. However, the proportion of investment in these high-end manufacturing industries was still at a relatively low level. In 2017, the share of general equipment manufacturing investment in total investment was 2.1%, a decrease of 0.1 percentage points over the previous year. The investment in special equipment manufacturing accounted for 2.0%, a decrease of 0.1 percentage points over the previous year; the investment in automobile manufacturing accounted for 2.1%, an increase of 0.1 percentage points over the previous year. The proportion of investment in railway, marine, aerospace and other transportation equipment manufacturing industries was 0.5%, which was the same as that of the previous year. The proportion of investment in computers, communications and other electronic equipment manufacturing was 2.0%, an increase of 0.3 percentage points over the previous year. It shows that although investment in high-end and equipment manufacturing industries had grown rapidly, these manufacturing investment accounts for less than 15% of total investment. Therefore, to fundamentally promote the upgrading of the manufacturing structure and improve the efficiency of industrial production, China must also effectively increase investment in high-end and equipment manufacturing.

Finally, from the perspective of sources of investment funds, in 2017, due to the implementation of a series of measures such as rectifying financial chaos, reducing leverage ratio, and preventing and controlling financial risks, fixed asset investment increased by 4.8%, which was a decrease of 1 percentage point over the previous year. Among them, the investment from domestic loans increased by 9.0%, a decrease of 0.9 percentage points over the previous year; the investment from other funds increased by 11.6%, a drop of 19.0 percentage points over the previous year; the recovery growth of corporate profits accelerated the growth of corporate investment to a certain extent, with investment from enterprises raising by 2.3%, an increase of 2.4 percentage points over the previous year. It is worth noting that in the past three years, the share of self-financing investment in total investment had continued declining: from 70.6% in 2015 to 66.7% in 2016, and further to 64.9% in 2017. The share of investment in other funds continued increasing: from 12.9% in 2015 to 16.1% in 2016, and further to 17.1% in 2017 (Fig. 1.9). In the situation where the share of self-financing investment of enterprises was still at a high level, the decline in the share of self-financing investment by enterprises could not be interpreted as a result of the improvement of the capital market; while the increase in the share of investment in other funds was an expression of the expansion of off-balance-sheet operations in the financial system.

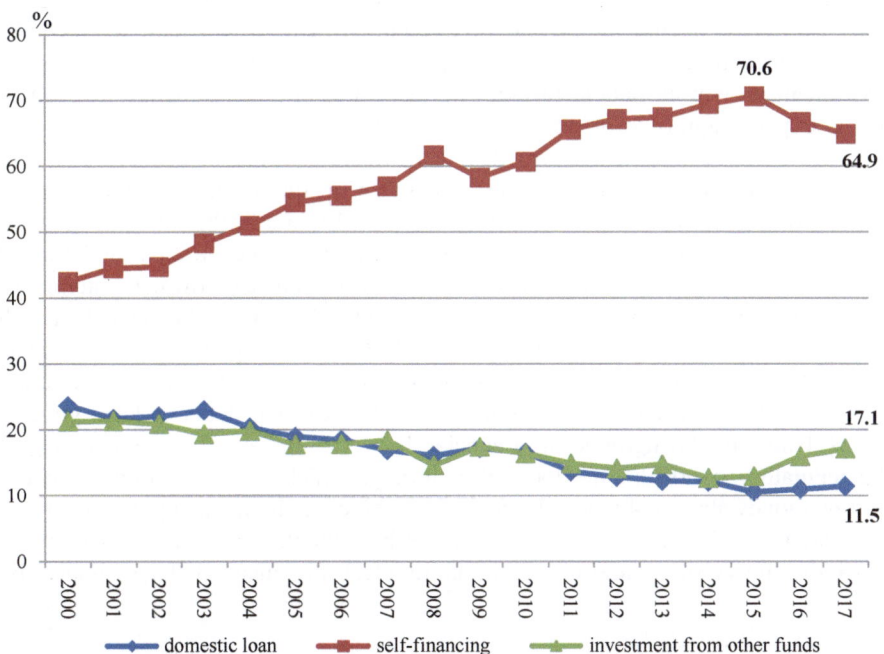

Fig. 1.9 Composition of source of investment in fixed assets investment. *Source* CEIC

In summary, the changes in the composition of the investment industry had obvious phased characteristics: After 2010, various measures of "steady growth" had substantially increased the share of investment in manufacturing and real estate; but after 2015, the share of investment in manufacturing had dropped sharply, although the share of investment in the real estate industry had declined, but it was relatively small. At the same time, the share of infrastructure investment had increased rapidly, from 18.4% in 2015 to 22.2% in 2017. It showed that if manufacturing investment was still an important force for economic growth from 2010 to 2014, then after 2015, its role began to give way to investment in real estate and infrastructure.

This shift corresponded to the change in private investment growth around 2015. After 2015, the slowdown in private investment growth was mainly attributed to the contraction of manufacturing investment, which in turn led to the decline in the growth rate of industrial added value. In order to offset this effect, the local governments had to increase its investment in infrastructure and in state-owned enterprises in the relevant monopoly areas (including the real estate sector). At the same time, the shrinking of the industrial sector directly restrained the growth of fiscal revenues of local governments and increased the dependence of local governments on revenue from land sales, thus resulting in the bubble in the real estate market. In this process, the financial discrimination against the private enterprises, the "implicit guarantees" of local governments in the capital market had led to the situation in which a large amount of capital was idling within the financial system without entering the real economy, increasing the possibility of financial systemic risk. The sharp decline in private investment growth, which had not received attention in a timely and effective manner, had led to the current shrinking in investment growth in manufacturing. The real estate and infrastructure investment had to expand rapidly to offset the decline in the total investment growth.

In fact, the decline in the growth rate of manufacturing investment in recent years was closely related to the continued sharp decline in private investment in manufacturing: from 27.2% in 2012 to 9.1% in 2015, and further to 3.6% in 2016, and to 4.8% in 2017. On the one hand, excessive concentration of financial resources in the real estate, infrastructure and in the state-owned enterprises, together with the existence of various investment barriers, had led to a weak growth in the manufacturing investment of the real economy, especially private enterprises.[1] On the other hand, relying on financial expansion to speed up infrastructure investment had also increased various hidden and explicit debts of local governments. Under the background of "financial deleveraging" and the strengthening of the regulatory oversight, the loopholes through which local governments could borrow money abnormally have basically been blocked. In the next phase, the rapid expansion of infrastructure investment might be difficult to sustain. In addition, in 2018, the supervision authorities continue to attack the financial chaos vigorously.

---

[1]Research team of Center for Macroeconomic Research at Xiamen University: "China's Macroeconomic Forecasting and Analysis—Spring 2016 Report".

While continuing to implement various real estate management and control policies, all kinds of violations will also be severely punished regarding the real estate
development loan. This will largely inhibit the growth of real estate investment.
Therefore, if the growth rate of private investment cannot rebound during the next
two years, the growth in investment in fixed assets will continue to slow down.

1.4   Industrial Production Growth Continues to Stabilize, and Profit Growth Between State-Owned and Private Enterprises Continues to Diverge

In 2017, the PMI was above "glory and withered line 50" for each month in the
year, and the PMI for export orders increased a lot. Export had played a support role
in maintaining the industrial growth. The annual industrial added value (with price
adjusted) increased by 6.6%, an increase of 0.6 percentage points over the previous
year. The stabilization of industrial production had been maintained, but the growth
rate was still at a low level since 2000 (Fig. 1.10). Since 2015, the growth rate of
industrial added value had started to fall below the growth rate of GDP. Although
the contribution of the secondary industry to GDP growth had dropped significantly
(Fig. 1.3), the stabilization of industrial production was still an important force for
stable economic growth.

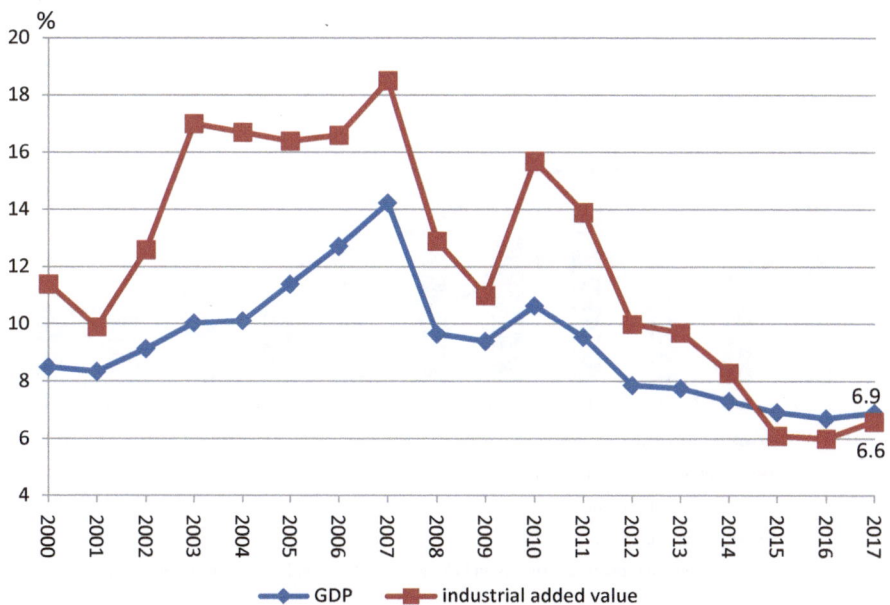

Fig. 1.10 Actual growth rate of real GDP growth and industrial growth. *Source* CEIC

From the perspective of industrial production of different ownership systems, the industrial added value of state-owned and state-controlled enterprises increased by 6.5% in 2017, an increase of 0.6 percentage points over the previous year; the industrial added value of the shareholding enterprises increased by 6.6%, a decrease of 0.3 percentage points over the previous year; the industrial added value of private enterprises grew by 5.9%, down by 1.6 percentage points over the previous year; the industrial added value of foreign investment enterprises and Hongkong, Macao and Taiwan investment enterprises grew by 6.9%, an increase of 2.4 percentage points over the previous year (Fig. 1.11). Since 2007, the growth rate of industrial added value of all types of ownership enterprises had shown a declining trend. Among them, the growth rate of joint-stock enterprises and private enterprises had fallen the most. After 2015, the growth rate of industrial added value of state-owned and state-controlled enterprise, of the foreign investment enterprises and of the Hongkong, Macao and Taiwan investment enterprises rebounded, but the trend of decline of joint-stock enterprises and private enterprises continued. It showed that the stabilization of industrial growth in recent years had benefited largely from the expansion of the production of state-owned and state-controlled enterprises.

At the same time, corporate profit growth had also started to stabilize and then picked up. In 2017, the total profit of industrial enterprises above designated size increased by 21%, an increase of 12.5 percentage points over the previous year. Among them, state-owned enterprises' profits increased by 39.9%, a significant increase of 41.1 percentage points over the previous year; state-controlled enterprises' profits increased by 45.1%, an increase of 38.4 percentage points over the

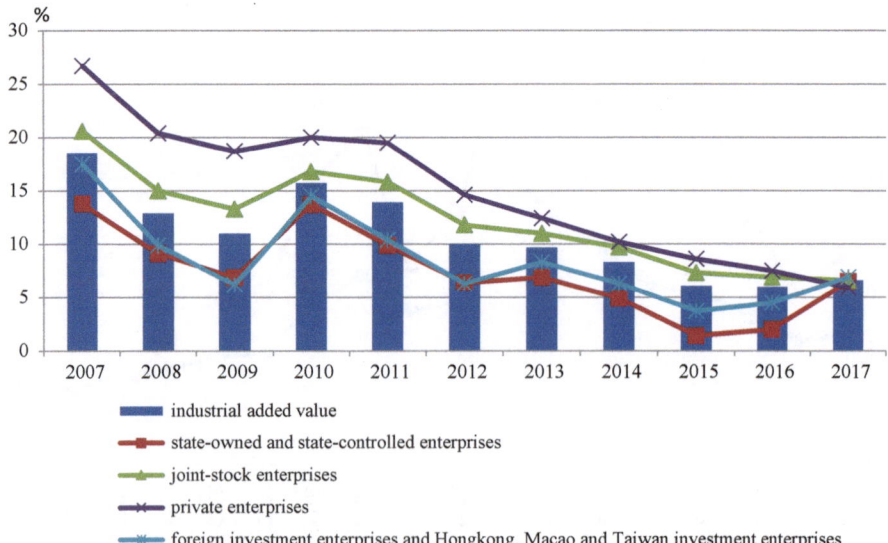

Fig. 1.11 The actual growth rate of industrial added value in different ownership enterprises. *Source* CEIC

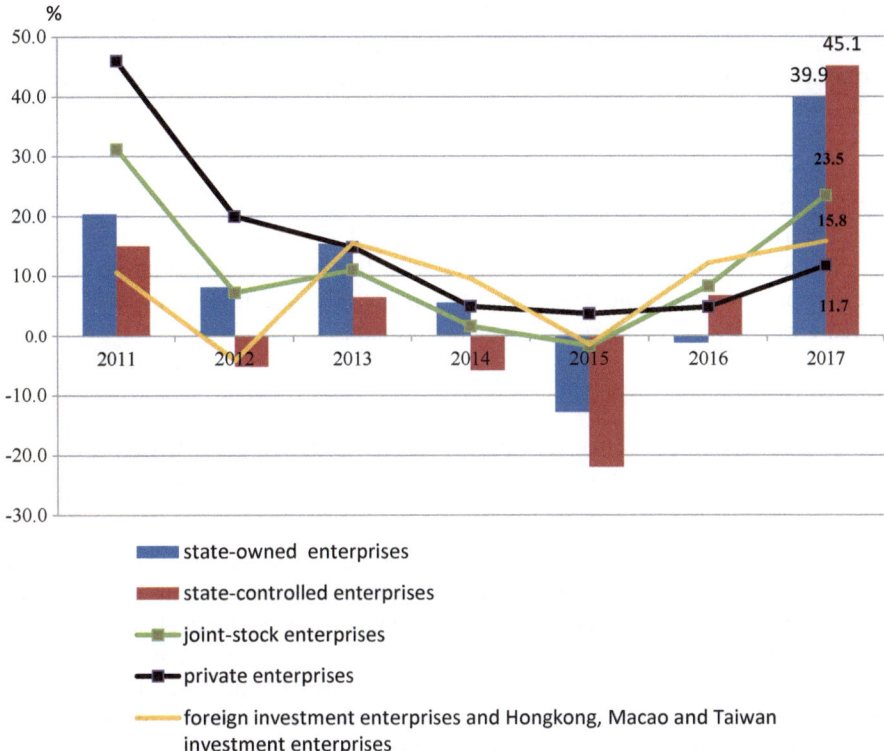

Fig. 1.12 Growth rate of total profit of different ownership enterprises. Source: CEIC

previous year; joint-stock enterprises' profits increased by 23.5%, an increase of 15.2 percentage points over the previous year. Profits of private enterprises increased by 11.7%, an increase of 6.9 percentage points over the previous year. Total profits of foreign investment enterprises and Hongkong, Macao and Taiwan investment enterprises increased by 15.8%, an increase of 3.7 percentage points over the previous year (Fig. 1.12). In addition, profits of large and medium-sized industrial enterprises increased by 27.0% in 2017, an increase of 17.1 percentage points over the previous year. Among them, the profits of the state-controlled enterprises increased by 49.6%, an increase of 42.6 percentage points over the previous year.

In fact, profit growth of private enterprises before 2015 was higher than that of state-owned and state-controlled enterprises. However, this trend was reversed: In the context of rising commodity prices and reduction of production overcapacity, the profit growth rate of upstream industries, the state-owned and state-controlled enterprises (especially large and medium-sized enterprises) in monopoly industries had rebounded sharply. However, the profit growth of the private enterprises in the competitive sector was relatively mild. The disparity of profit growth between state-owned and private enterprises in the short term had ensured the growth of tax revenue, it had long been detrimental to the improvement of industrial production efficiency and the upgrading of industrial structure. More importantly, it would further reduce the allocation efficiency of resources for production factors among industries.

Under the influence of various measures such as "deleveraging", the asset-liability ratio of industrial enterprises in 2017 was 55.5%, which was basically the same as that of the previous year. Among them, the highest debt-equity ratio of state-controlled enterprises was 60.4%, and the debt-to-equity ratio of joint-stock enterprises was 56.2%, a decrease of 1.2 and 0.3 percentage points respectively over the previous year; the debt-to-equity ratio of private enterprises was 51.6%, an increase of 0.9 percentage points over the previous year.

1.5   The Growth Rate of Import and Export Rebounded and the RMB Exchange Rate Continued to Rise

In 2017, the recovery of the world economy had effectively promoted the growth of China's foreign trade. Under the effect of the rebound of the price indices of import and export, the total import and export volume had reversed the decline trend in the past two years. Among them, exports (RMB) increased by 10.8%, an increase 12.8 percentage points over the previous year; imports (RMB) increased by 18.7%, an increase of 18.1 percentage points over the previous year. Balance of imports and exports, trade surplus reached 2871.6 billion yuan, a decrease of nearly 473.4 billion yuan over the previous year (Fig. 1.13).[2]

The trade structure continued to adjust, and the share of general trade continued to increase in 2017. Among them, the share of general trade exports in total exports accounted for 54.5%, an increase of 1.7 percentage points over the previous year; the export ratio of processing trade accounted for 34.0%, a decrease of 1 percentage point over the previous year. General trade imports accounted for 57.4% of total imports, up 1.9 percentage points over the previous year; processing trade imports accounted for 23.9%, down 0.8 percentage points over the previous year.

From the perspective of the ownership structure of import and export companies in 2017, the share of total exports of private enterprises in exports accounted for 44.4%, exceeding the ratio of 43.2% of foreign-invested enterprises for the first time; the share of state-owned enterprises accounted for 10.2%. Foreign-invested enterprises accounted for 38.1% of total imports. Import of private enterprises accounted for 22.1% of total import, and state-owned enterprises accounted for 23.8%. It shows that private enterprises played an important role in total export growth.

---

[2]In terms of dollar, exports grew by 7.9% in 2017, an increase of 15.6 percentage points over the previous year; imports increased by 15.9%, and the growth rate accelerated by 21.4 percentage points from the previous year. The trade surplus reached 422.54 billion U.S. dollars, reduced by 87.16 billion yuan over the previous year.

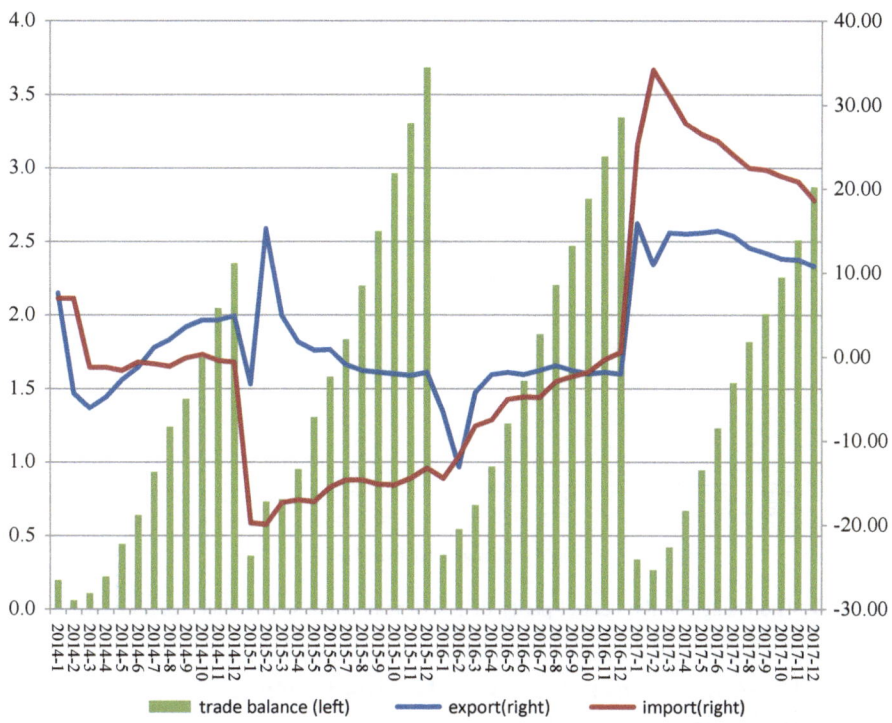

Fig. 1.13 Accumulated growth rate of imports and exports (RMB value) and changes in trade balance. *Source* CEIC

Since the central bank initiated the reform of the RMB central parity pricing mechanism in August 2015, the unilateral appreciation of the RMB against the US dollar had been effectively reversed. The devaluation of the RMB against the US dollar in the onshore market had effectively inhibited the capital outflow of China. In 2016, China's foreign exchange reserved totaled 3.01 trillion US dollars. In 2017, the decline of the U.S. dollar index prompted the RMB to start to appreciate against the US dollar. By the fourth quarter of 2017, the median price of U.S. dollar against the U.S. dollar was 6.609 yuan, and the annual appreciation of the U.S. dollar against the U.S. dollar had risen by 3.3% (Fig. 1.14). Foreign exchange had reserved totaled 3.14 trillion U.S. dollars. As a whole, the exchange rate of the RMB against the US dollar began to go out of a "bilateral floating, basically stable" situation.

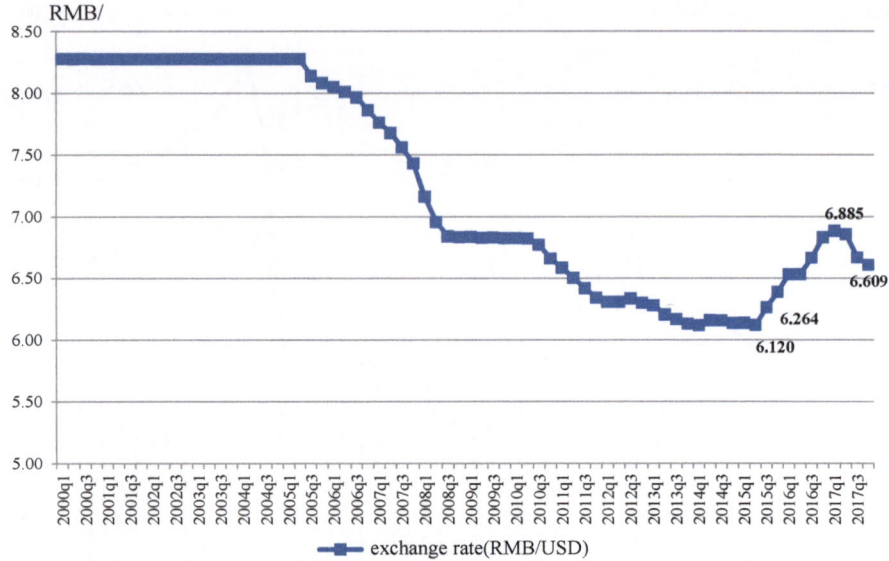

Fig. 1.14 The change in the central parity of RMB against the U.S. dollar. *Source* CEIC

1.6 CPI Rises Moderately, PPI Begins to Rise, and Residents' Real Income Increases Steadily

In 2017, the Consumer Price Index (CPI) continued its modest upward trend, rising by 1.6%, which was a 0.4 percentage points drop from the previous year. Affected by the rise in international commodity prices and reduction of overcapacity, plus the base effect, the industrial Producer Price Index (PPI) rose by 6.3%, ending the five consecutive years of decline since 2012 (Fig. 1.15). The impact of the PPI rebound on CPI remains to be gradually released.

In 2017, residents' income had achieved steady growth. The per capita disposable income of residents in the country increased by 7.3%, an increase of 1.0 percentage points over the previous year but a slight decrease of 0.1 percentage point over 2015. Among them, the per-capita real disposable income of urban residents increased by 6.5%, and the per capita disposable income of rural residents increased by 7.3%, which increased by 0.9 and 1.1 percentage points respectively over the previous year, but decreased by 0.1 and 0.2 percentage points respectively over 2015 (Fig. 1.16). The total retail sales of consumer goods reached a nominal growth of 10.2%, a slight decrease of 0.2 percentage points over the previous year. With price adjusted, per capita consumer spending increased by 5.4%, which was a 1.4 percentage points decrease over the previous year. Among them, the per capita consumption expenditure of urban residents increased by 4.1%, and the growth rate was down by 1.6 percentage points over the previous year; the per capita consumer spending of rural residents increased by 6.8%, a decrease of 1.0 percentage points

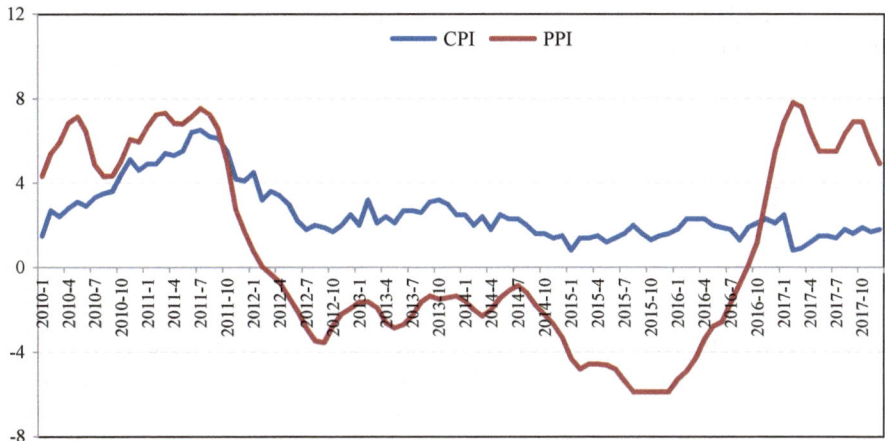

Fig. 1.15 Changes in major price indices. *Source* CEIC

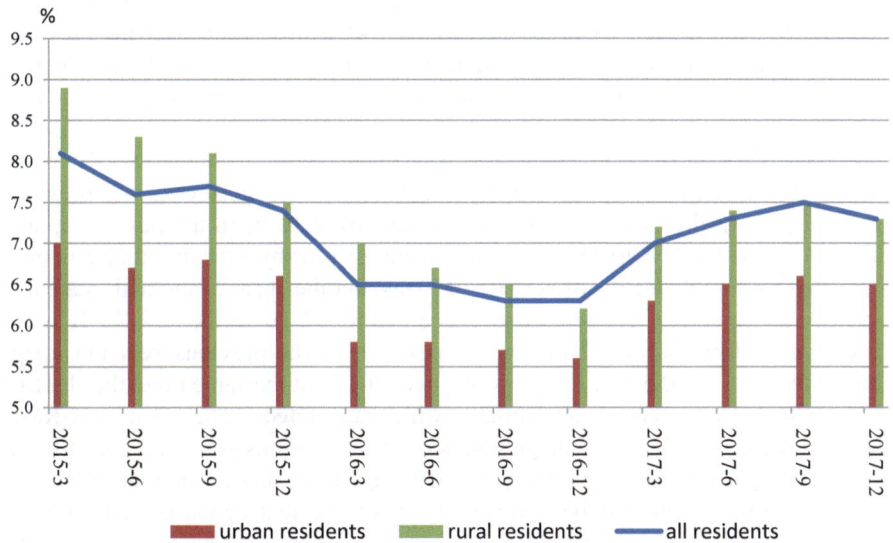

Fig. 1.16 Changes in the growth rate of actual disposable income of residents. *Source* CEIC

over the previous year. In recent years, the slowdown in the growth rate of labor productivity had been the main reason for curbing the rapid increase in real income of residents. In the foreseeable future, it is unlikely that the share of secondary industry in China will decline rapidly. Then, the rapid growth of labor productivity in industry, especially in the manufacturing industry, must be an important guarantee for economic labor productivity and the rapid increase in the real income growth of residents. For this reason, the rapid expansion of manufacturing private investment is indispensable.[3]

1.7 The Overall Tightening of Monetary Policy and the Composition of Financial Resources Need to Be Adjusted

In the second half of 2017, the chaos of the financial system began to be rectified, which significantly curbed the growth of money supply. The annual cash flow (M0) increased by 3.4%, M1 increased by 11.8%, M2 increased by 8.2%, and the growth rate dropped by 4.7, 9.6 and 3.1 percentage points respectively over the previous year (Fig. 1.17). In 2017, the M2 growth declined owing to two reasons: the strengthening of financial supervision which weakened the ability of banks to derive credit through non-bank channels, and the shrinking of deposits.[4]

In 2017, the total amount of new social financing was 19.44 trillion yuan, an increase of 1.64 trillion yuan over the previous year. Among them, the new RMB loans were 13.84 trillion yuan, an increase of 1.41 trillion yuan over the previous year. The proportion of newly added RMB loans in total social financing was 71.2% for the year, an increase of 1.34 percentage points over the previous year (Fig. 1.18).

In 2017, then on-financial enterprises and organizations accounted for 49.6% of the new RMB loans, which was 1.4 percentage points higher over the previous year, but a sharp drop of 13.4 percentage points over 2015. In the past two years, although the new RMB loans had been expanding year by year, the real economy accounted for less than 50% of new RMB loans, but the share of the real estate had increased substantially. The real estate accounted for 41.1% of the new RMB loans, which was a decrease of 3.7 percentage points over the previous year but a significant increase of 10.5 percentage points over 2015. At the same time, the share of residents had also increased substantially, which accounted for 52.7% of new RMB loans, an increase of 2.7 percentage points over the previous year, and a significant increase of 19.7 percentage points over 2015 (Fig. 1.19). It showed that in the past two years, investment had been diverted out of the real economy, and a large amount of capital was idling within the financial system without entering the real economy. This not only exacerbates the bubble of the real estate market, but also exacerbates the risks of the financial system. The adjustment of the composition of financial resources had become an urgent task.

---

[3]Research team of Center for Macroeconomic Research, Xiamen University: "China's Macroeconomic Forecasting and Analysis—Fall 2017 Report."
[4]See http://www.guancha.cn/Wanzhao/2018_02_01_445400.shtml.

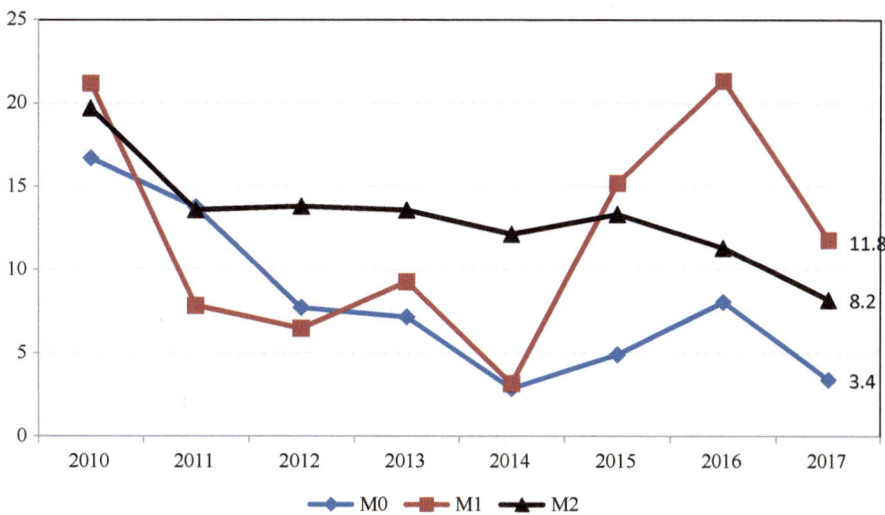

Fig. 1.17 Growth rate of money supply. *Source* CEIC

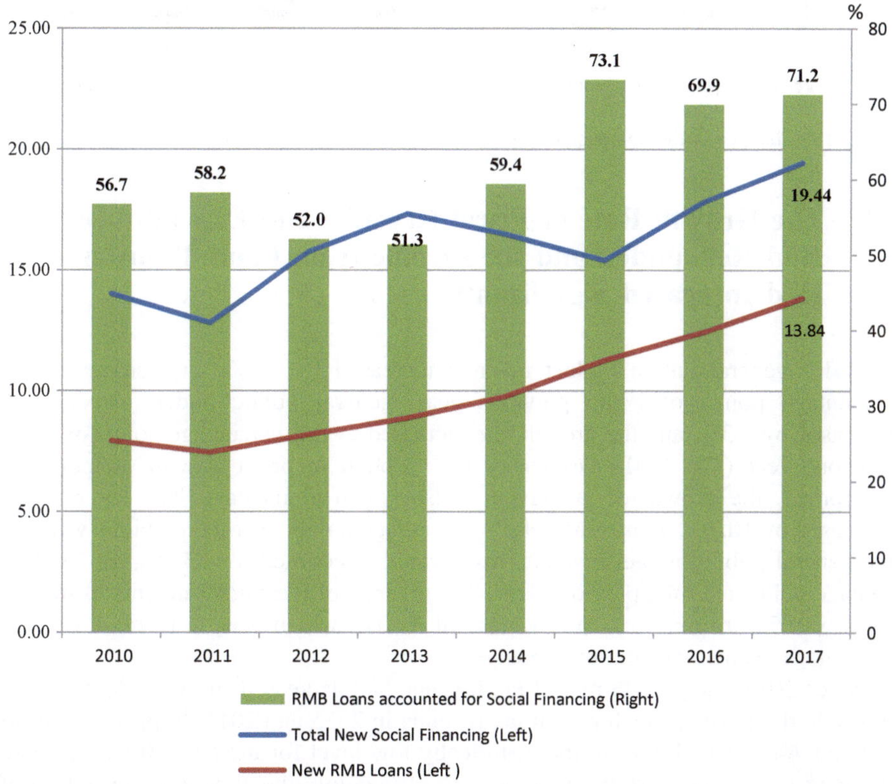

Fig. 1.18 New increase in social financing and total RMB loans. *Source* CEIC

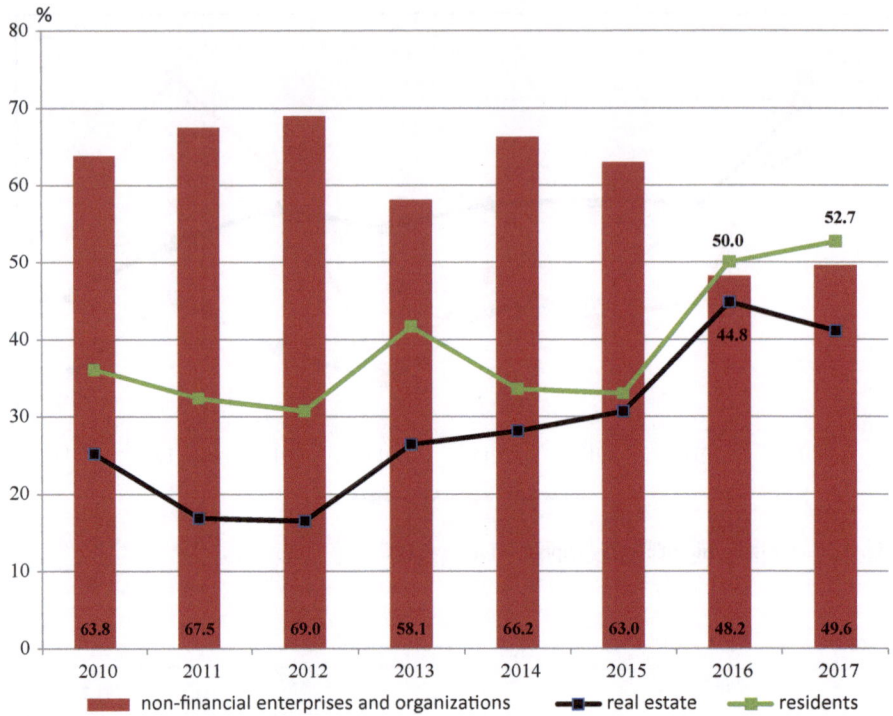

Fig. 1.19 Changes in the composition of new RMB loans. *Source* CEIC

1.8 The Growth Rate of Fiscal Revenue and Expenditure Had Rebounded and the Income from Land Transfer Had Increased Significantly

In 2017, general public budget revenue increased by 8.1%, an increase of 3.3 percentage points over the previous year; general public budget expenditure increased by 8.3%, and the growth rate increased by 1.5 percentage points over the previous year (Fig. 1.20). Driven by the rebound in profit growth of industrial enterprises, the growth of tax revenue also began to increase. The tax revenue increased by 10.7%, an increase of 6.4 percentage points over the previous year. Of the general public budget revenue, tax revenue accounted for 83.7%, up 2.0 percentage points over the previous year. The revenue of local governments from land sales was 5.2 trillion yuan, an increase of 39.0%, which was an increase of 23.9 percentage points over the previous year.

Since 2013, the growth rate of tax revenue had decreased from the double-digit to single-digit. The growth rate of tax revenue in 2015 and 2016 dropped further to 4.8 and 4.4%, which was at the historically low level for the past 30 years. Since the industrial taxes were the main source of taxation, the slowdown in industrial

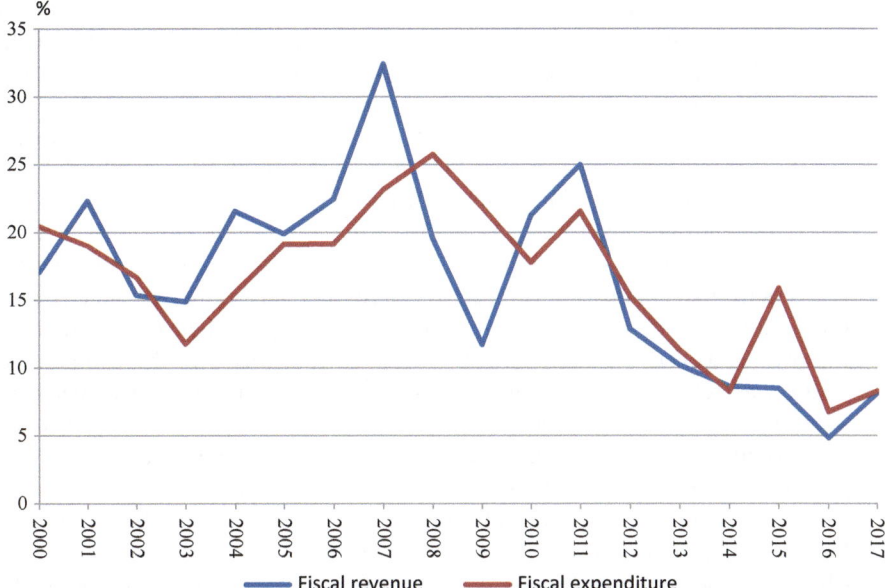

Fig. 1.20 Changes in the growth rate of income and expenses of the general public budget. *Source* CEIC

growth and the decline in the growth rate of corporate profits have greatly inhibited the growth of tax revenues. In order to ensure fiscal revenues, local governments not only tended to drive up the land price by restricting land supply, but also tended to obtain loan through land mortgage. This had directly increased the debt burden of local governments, which in turn had also increased the risk of the financial system.

In summary, although China's economic growth maintained stable under the positive influence of world demand in 2017, the features that have emerged since 2015 indicate that the factors behind the promotion of the economic growth have become more complex. The prospects for economic growth may also be more uncertain. The following six issues need to be given full attention:

(a) While the share of the secondary industry is basically the same, the contribution rate of the industrial sector to GDP growth had dropped significantly, which, to a large extent, reflected the low efficiency of industrial production and the need for further advancement of industrial structure adjustment.

(b) The growth rate of investment in fixed assets continued to fall, and the investment structure was further out of balance. In the aspect of ownership structure, the share of state-owned enterprises' investment continued to rise, and the share of private investment continued to decline; in terms of industry structure, the share of manufacturing investment continued to decline, the share of infrastructure investment continued to rise, and the share of real estate investment remained high. This shows that the investment structure had not been adjusted in the direction of improving investment efficiency. The next phase of economic growth will not only be constrained by investment growth, but the decline in investment efficiency will also damage economic efficiency and increase the fragility of the financial system.

(c) The stabilization of industrial production had been maintained in recent years, but the growth rate was still at a low level since 2000. The stabilization of industrial production benefited largely from the expansion of the state-owned and state-controlled enterprises. With the increase in commodity prices and the reduction of production overcapacity, the profit growth rate of state-owned and state-controlled enterprises (especially large and medium-sized enterprises) involved in upstream production materials and monopoly industries rebounded sharply. However, the profit growth of the private sector in the competitive sector was relatively mild. Although the disparity of profit growth between state-owned and private enterprises in the short term had ensured the growth of tax revenue, it had long been detrimental to the improvement of industrial production efficiency and the upgrading of industrial structure. More importantly, it will further reduce the allocation efficiency of resources for production factors among industries.

(d) China must prevent investment from being diverted out of the real economy. In the past two years, although the new RMB loans had been expanding year by year, the real economy had accounted for less than 50% of the new RMB loans. At the same time, the share of loans to real estate and to the resident sector had increased significantly. The non-financial companies and organizations accounted for 49.6% of the new RMB loans, which was 1.4 percentage points higher than the previous year, but a sharp drop of 13.4 percentage points over 2015. The resident sector accounted for 52.7%, which was not only 2.7 percentage points higher than the previous year, but also a significant increase of 19.7 percentage points over 2015. Under the condition of implicit government guarantees, the funds released by the loose monetary policy strongly support the real estate enterprises, local financing platforms, and the zombie enterprises through the shadow banking system, which in turn drove up the leverage ratio of state-owned enterprises and the local governments. As a result, the allocation of financial resources to the real estate industry, infrastructure, and state-owned enterprises with excessive concentration, together with the existence of various investment barriers, had led to a sluggish of investment growth in the real economy, especially private enterprises. As a result, the high leverage ratios of the financial sector and state-owned enterprises have emerged, becoming the two major financial risk points for state-owned companies in the financial and non-financial sectors. Therefore, the adjustment of the composition of financial resources had become an urgent task.[5]

---

[5]In January 2018, RMB loans increased by 2.9 trillion yuan, an increase of 877 billion yuan year-on-year. Among them, loans from non-financial companies and organizations increased by 1.78 trillion yuan, accounting for 61.4% of the total. Although the scale of loans to the real economy had expanded by 220 billion yuan over the previous year, its share had dropped by 15.5

(e) The non-government investment could not rebound quickly since 2015, and this problem had not been fully appreciated. First, the decline in the growth rate of non-government investment was reflected in the shrinking of investment in manufacturing, which in turn led to the increase in industrial value and the decline in the growth rate of corporate profits. Second, to counteract this effect, the government had to expand its investment in infrastructure and state-owned enterprises in some monopolist areas (including the real estate sector). Third, the slowdown of profit growth in the industrial sector directly restrained the growth of fiscal revenues of local governments, thereby increasing their dependence on revenue from land sales, which boosted the bubble in the real estate market. Finally, owing to the long-term tilt of loan policy favorable toward state-owned enterprises and the soft budget constraints of state-owned enterprises, the huge liquidity injected into the economy for steady growth had not only rapidly raised the high debt of non-financial state-owned enterprises, but also has further hindered the rebound of private investment.[6]

(f) Although the residents' income grew steadily in 2017, the decline in labor productivity growth was a fundamental reason curbing the rapid increase in real income of residents. In the future, the rapid growth of labor productivity, especially the manufacturing industry, will inevitably be an important guarantee for the rapid increase in the real income of residents. Although investment in high-end and equipment manufacturing industries had grown rapidly in recent years, these manufacturing investment accounted for less than 15% of total investment. To fundamentally promote and upgrade the manufacturing structure and improve the efficiency of industrial production, China must effectively increase investment in high-end and equipment manufacturing. Among them, the rapid expansion of manufacturing private investment is indispensable

With forty years of reform and opening up, China's economy had achieved brilliant results. The scale of GDP in 2017 was 32.3 times larger than in 1978, and the level of per capita GDP reached 22.4 times that of 1978. With the continuous expansion of the economic scale and continuous improvement of the people's income level. The main contradiction in social development is transformed from "the contradiction between the people's ever-increasing material and cultural needs and backward social production" to the "contradictory between the people's growing good living needs and unbalanced development." To resolve this major contradiction, the economy and society need to be transformed from the pursuit of "high speed growth" to the pursuit of "high quality development." To this end, China must face up to and solve the various types of "unbalanced development". Among them, the imbalance between the development of the real economy and the

percentage points from the same period of last year. In 2018, the scale of domestic loans is still greatly expanding, and the proportion of loans to the real economy is still falling sharply.

[6]Research team of Center for Macroeconomics Research, Xiamen University: "China's Macroeconomic Forecasting and Analysis—Fall 2017 Report".

virtual economy is reflected in the insufficient financial service for real economy, and it had become one of the main reasons for the expansion of financial risks.

In the next two years, on the one hand, under the background of "financial deleveraging" and the strengthening of the regulatory oversight of local finances, the loopholes through which local governments could borrow money abnormally will basically be blocked, and the rapid expansion of infrastructure investment will be difficult. In order to succeed, on the other hand, after entering 2018, the supervision department will continue to strike out and vigorously rectify all kinds of financial chaos. While continuing to implement various real estate management and control policies, various types of irregularities surrounding real estate development loans will also be severely punished. This will largely inhibit the growth of real estate investment. Therefore, if the growth rate of private investment cannot rapidly rebound during the next two years, the growth of investment in fixed assets will continue to slow down, and the economic growth will also be difficult to rise. More importantly, with the economic slowdown, local governments demand for infrastructure investment and their dependence on revenue from land sales, the financial sector's preference for state-owned enterprises and the real estate industry will hardly be changed. China had to prevent investment from being diverted out of the real economy, in order to reverse the situation in which a large amount of capital is idling within the financial system without entering the real economy. Otherwise, the two major financial risks of financial and non-financial state-owned enterprises will also be difficult to fundamentally eliminate.

Based on this, the policy simulation of this report will focus on the quantitative analysis of the main tasks that drive state-owned enterprises to reduce leverage ratio, stimulate private investment growth, stabilize economic growth, and achieve high-quality development, and the report will discuss how to promote the rapid rebound of private investment growth.

Chapter 2
China's Macroeconomic Forecast for 2018–2019

2.1 Assumptions of Model Exogenous Variables

2.1.1 Economic Growth Rate in the United States and the Eurozone

In 2017, the world economies recovered strongly. The U.S. economy grew by 2.3%, an increase of 0.7 percentage points over the previous year; the Eurozone's economy grew by 2.5%, and the growth rate reached the highest level since 2007. In the next two years, although the U.S. economy is still facing greater risks, the release of tax reform effects, stable domestic consumption growth, and the weak US dollar policy will all help maintain the steady growth of U.S. economy. The International Monetary Fund (IMF) released its forecast in January 2018. In the next two years, the growth rate of the US economy is expected to remain at 2.7 and 2.5%. Based on this, the research team sets the growth rate of the U.S. economy in the next eight quarters (Fig. 2.1).

On the other hand, with the disappearance of the uncertainty of the European debt crisis and the ease of the concerns that the Eurozone will split, it is expected that the Eurozone economy will continue to maintain strong growth. The IMF forecast in January 2018 that the economic growth rate in the Eurozone would reach 2.2 and 2.0% respectively in the next two years. Taking into account that the previous IMF's forecast of economic growth in the Eurozone in 2017 (2.4%) was lower than the actual value (2.5%), its forecasts for 2018 and 2019 may also be underestimated. For this reason, the research team assumes that the economic growth rate of the Eurozone in 2018 and 2019 will reach 2.3 and 2.1%, respectively, which is 0.1 percentage points higher than the IMF forecast of January 2018. The quarterly year-on-year growth rate setting is based on the growth trend of the Eurozone economy in the past and is set at "low at the beginning and then high" and will reach the highest value in the third quarter of 2018. Subsequently, as the loose monetary policy may withdraw, the economic growth rate gradually rebound (Fig. 2.1).

© Springer Nature Singapore Pte Ltd. 2018
Center for Macroeconomic Research at Xiamen University, *China's Macroeconomic Outlook*, Current Chinese Economic Report Series,
https://doi.org/10.1007/978-981-13-1005-8_2

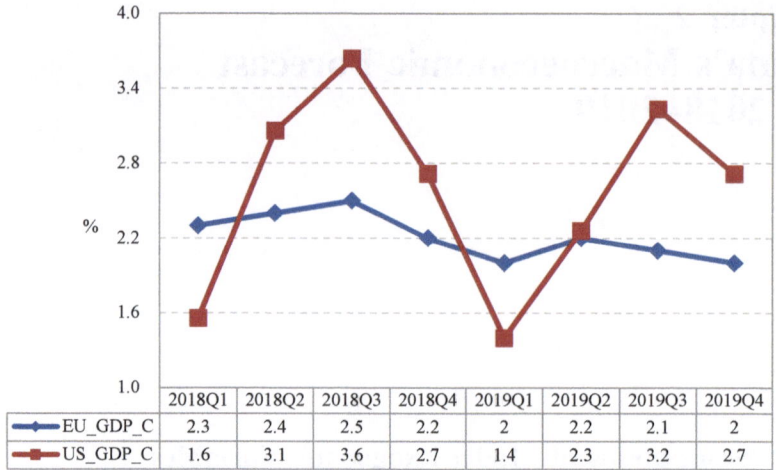

	2018Q1	2018Q2	2018Q3	2018Q4	2019Q1	2019Q2	2019Q3	2019Q4
EU_GDP_C	2.3	2.4	2.5	2.2	2	2.2	2.1	2
US_GDP_C	1.6	3.1	3.6	2.7	1.4	2.3	3.2	2.7

Note: 1. EU_GDP_C represents the growth rate of real GDP in the Eurozone,
and US_GDP_C represents the growth rate of real GDP in the United States;
2. Eurozone data are seasonally adjusted year-on-year growth rates;
U.S. data are converted to annual quarter-to-quarter annualized rates.

Fig. 2.1 Assumptions of changes in the economic growth rate of the United States and the euro. *Source* Assumption by research team. Note: (1) EU_GDP_C represents the growth rate of real GDP in the Eurozone, and US_GDP_C represents the growth rate of real GDP in the United States; (2) Eurozone data are seasonally adjusted year-on-year growth rates; U.S. data are converted to annual quarter-to-quarter annualized rates

2.1.2 The Main Exchange Rates

The better-than-expected recovery of the Eurozone's and Japan's economy in 2017 led to an expectation that the monetary policy of these economies may contract and formed pressure of the depreciation of the U.S. dollar. At the same time, the U.S. government's short-term weak US dollar policy further promoted the decline of the US dollar index.[1] Affected by this, in 2018, the exchange rate of RMB against the US dollar begins to appreciate rapidly. However, taking into account that the US monetary policy may turn to tightening in 2018, the Fed will continue to raise interest rates during the year and to reduce its balance sheet at the same time. As a result, the US dollar index has a limited space for decline. On the other hand, taking into account the various risks, the foundation for the continued recovery of China's economic growth the next two years is not solid, so the possibility that the RMB will continue to appreciate sharply against the US dollar is not high. The research team assumes that the RMB exchange rate against the U.S. dollar will close at 6.46 in 2018: In the first quarter, the U.S. dollar exchange rate will continue to remain

---

[1]On February 1, 2018, the U.S. dollar index fell to 88.5, returning to the level of the end of 2014.

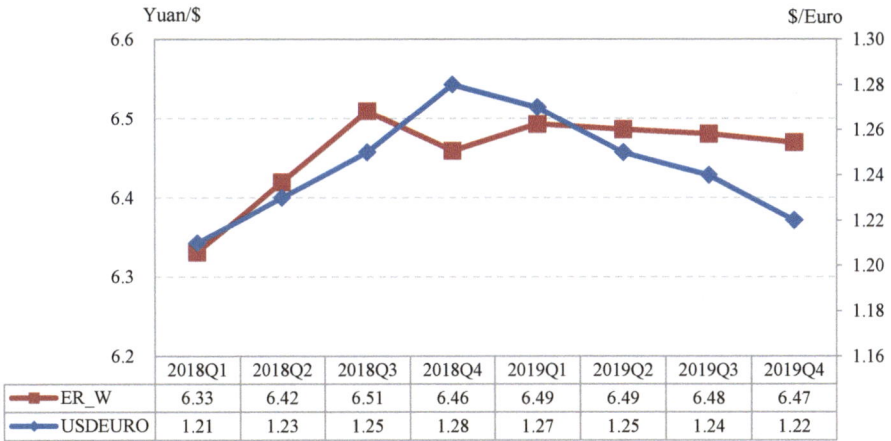

	2018Q1	2018Q2	2018Q3	2018Q4	2019Q1	2019Q2	2019Q3	2019Q4
ER_W	6.33	6.42	6.51	6.46	6.49	6.49	6.48	6.47
USDEURO	1.21	1.23	1.25	1.28	1.27	1.25	1.24	1.22

Fig. 2.2 Assumed changes in the exchange rate of the US dollar against the Euro and the exchange rate of the RMB against the U.S. dollar. *Source* Assumption by research team

below 6.40, and it will likely recover in the second and third quarters. In the fourth quarter, affected by the mid-term U.S. government elections, the expectations of appreciation of the RMB may start again. In 2019, the RMB exchange rate will be basically stable, with the overall level approaching 6.50 (Fig. 2.2).

As for the exchange rate of the euro against the US dollar, considering that the momentum of economic growth in the Eurozone may be stronger than that of the United States in the next two years, it is more likely that the euro will continue to appreciate against the US dollar in the short term. The research team assumes that the exchange rate of the euro against the US dollar will maintain a steady rise in 2018, and that by the end of the year, the exchange rate will be 1.28 US dollars per euro; by the end of 2019, it will fall to 1.22 US dollars (Fig. 2.2).

2.1.3 Growth Rate of Broad Money Supply (M2)

In 2017, the growth rate of China's broad money supply M2 was 8.2%, which was significantly lower than the target growth rate of 12% at the beginning of the year. The monetary policy in 2018 is expected to continue to maintain "steadier and tighter": First, the monetary authorities will continue to maintain strict supervision and financial deleveraging, prevent and control the impact of changes in the market environment on the financial system, avoid the risk of shadow banking and asset price adjustments, and prevent "Minsky Moment" from happening; second, the price adjustment instruments tend to tighten, and the interest rate levels of MLF, SLF, reverse repurchase and other operations will fluctuate as a whole. This not only fits the upward trend of the international interest rate, but also narrows the disparity between the operating interest rate and the money market interest rate,

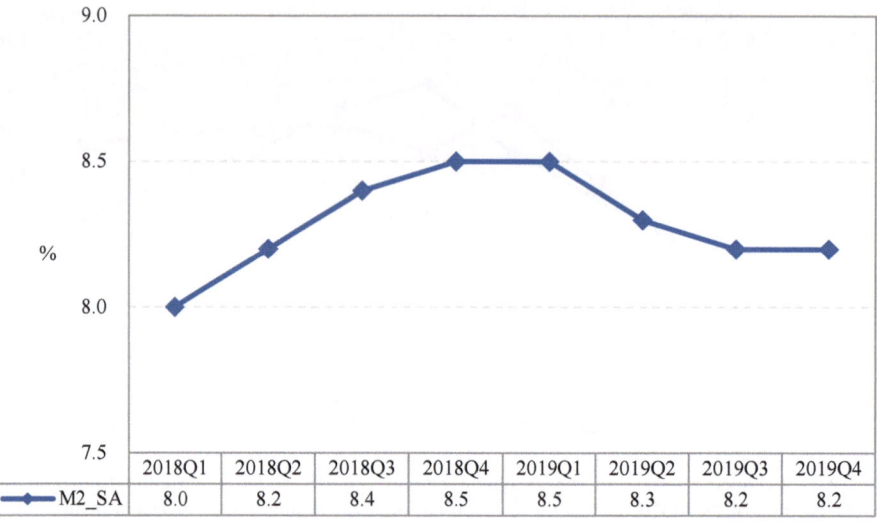

	2018Q1	2018Q2	2018Q3	2018Q4	2019Q1	2019Q2	2019Q3	2019Q4
M2_SA	8.0	8.2	8.4	8.5	8.5	8.3	8.2	8.2

Fig. 2.3 M2 growth rate trend assumption. *Source* Assumption by research team

maintains the Sino-U.S. spread over the safety boundary, and restrains the market from excessive speculation or arbitrage; finally, the quantitative tool maintains moderate growth. It uses open market operations and innovative financial instruments to proactively regulate liquidity. While satisfying market liquidity requirements, to support the development of the real economy, China must prevent market interest rates from rising excessively. Therefore, the research team assumes that M2 is expected to grow by 8.5% in 2018, a slight increase of only 0.3 percentage points over the previous year; in 2019, structural optimization adjustments will be brought about by structural reforms in the domestic supply side and financial deleveraging. The increase in the money supply will return to 8.2% (Fig. 2.3).

2.2   Prediction of China's Main Macroeconomic Indicators for 2018–2019

2.2.1   GDP Growth Forecast

Based on the forecast results of the China Quarterly Macroeconomic Model (CQMM), China's economic growth will continue to "stabilize and moderate" in 2018, and GDP is expected to increase by 6.73%, a decline of 0.17 percentage points over the previous year. The GDP growth rate will further decline to 6.60% in 2019. In the next two years, the economy will still maintain a growth rate of over 6.5%, and the possibility of a sharp drop is very low.

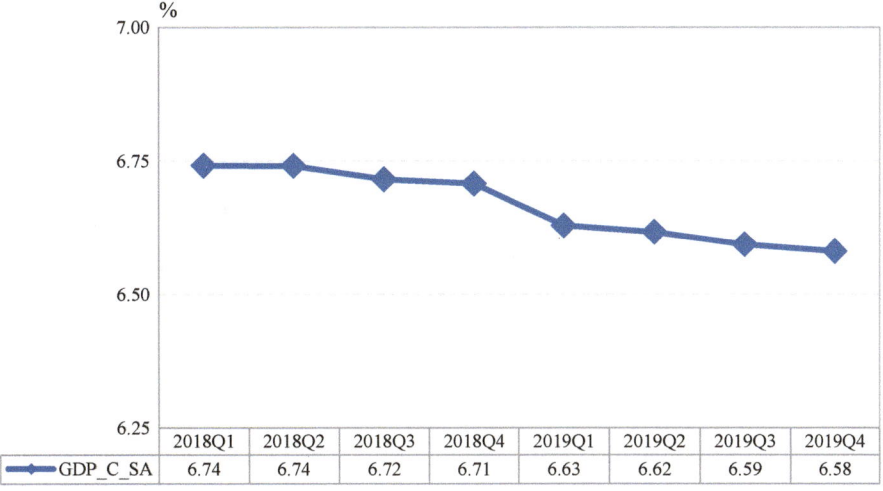

Fig. 2.4 Quarterly GDP growth forecast (quarterly YoY growth rate). *Source* Calculations by research team

From the quarterly growth rate, due to financial deleveraging and the continuous deceleration of investment, the GDP growth rate in the four quarters of 2018 is expected to reach 6.74, 6.74, 6.72, and 6.71%, respectively. With the slowdown of growth rate of the world economy and further adjustments of the domestic economic structure, it is expected that the growth rate of GDP of China will be 6.63, 6.62, 6.59 and 6.58% in the four quarters of 2019 (Fig. 2.4).

2.2.2   Major Price Index Forecast

The research team forecasts that CPI will increase by 2.13% in 2018, an increase of 0.53 percentage points over the previous year. By 2019, due to the continuously tightening monetary environment, the CPI is expected to fall back to 1.94%. On the other hand, commodity prices are expected to stabilize at high levels in the next two years, but the producer price index (PPI) will fall to a greater extent due to the big base: PPI is expected to rise by 4.64% in 2018, and by 1.99% in 2019. In 2018, the GDP deflator (P_GDP) will increase by 3.66%, a decrease of 0.45 percentage points from 2017, and may further decline to 2.80% in 2019. The forecast results show that the inflation level of China's economy can still be kept within a controllable range in the next two years.

Quarterly, after the second quarter of 2018, the increase in PPI over CPI will quickly shrink; after the third quarter of 2019, the increase in PPI will be smaller than the increase in CPI. In general, the price index basically shows the trend of "first increase and then decrease" (Fig. 2.5).

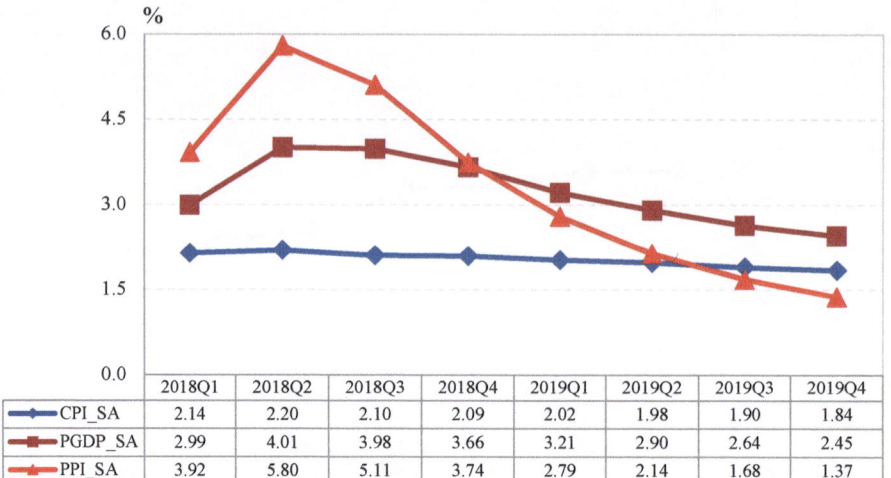

	2018Q1	2018Q2	2018Q3	2018Q4	2019Q1	2019Q2	2019Q3	2019Q4
CPI_SA	2.14	2.20	2.10	2.09	2.02	1.98	1.90	1.84
PGDP_SA	2.99	4.01	3.98	3.66	3.21	2.90	2.64	2.45
PPI_SA	3.92	5.80	5.11	3.74	2.79	2.14	1.68	1.37

Note: CPI_SA, PGDP_SA and PPI_SA represent quarterly adjusted consumer price index,

GDP deflator and producer price index, respectively

Fig. 2.5 Forecast of price index (quarterly YoY growth rate). *Source* Calculations by the research team. Note: CPI_SA, PGDP_SA and PPI_SA represent quarterly adjusted consumer price index, GDP deflator and producer price index, respectively

2.2.3 Forecast of Other Major Macroeconomic Indicators Growth

(a) Import and Export Growth Rate and Foreign Exchange Reserve Forecast

Beneficial from the better-than-expected growth of the world economy, the export growth of China is forecasted to maintain steady growth in the next two years. The total export value (current US dollar value) will increase by 9.65% in 2018, an increase of 1.75 percentage points over the previous year. In 2019, the total export value (current US dollar value) will increase by 9.35%, a slight decrease of 0.3 percentage points over 2018. From a quarterly perspective, the year-on-year growth of exports in 2018 showed a trend of "low first and then high", or the growth rate in the second half of the year will be significantly faster than the first half of the year. On the other hand, the slowdown in the growth rate of the domestic economy in the next two years may inhibit the growth of imports. In 2018, the total value of imports (current US dollar value) is expected to increase by 12.32%, a decrease of 3.58 percentage points over the previous year; the growth rate of total imports (current US dollar value) in 2019 will likely further fall back to 9.64% (Table 2.1).

In 2018, the proportion of net exports of goods and services to GDP in China will drop to 1.10%, a decrease of 0.44 percentage points over the previous year; in 2019, it will rise back to 1.13%. The stabilization of the RMB exchange rate against

Table 2.1 Forecast of China's import and export growth and net export as a share of GDP from 2018 to 2019 (unit: %)

Time	Export				Import				Net exports account for The proportion of GDP
	Constant price RMB	Current price USD	General trade Current price/USD	Processing trade Current price/USD	Constant price RMB	Current price USD	General trade Current price/USD	Processing trade Current price/USD	
2018	10.74	9.65	12.67	5.15	9.67	12.32	15.15	5.03	1.10
Q1	9.69	8.68	14.07	2.67	4.45	10.47	12.71	1.92	1.15
Q2	6.68	6.54	11.43	2.96	7.56	11.94	16.07	1.66	1.03
Q3	12.85	12.14	14.26	7.55	11.74	13.82	17.97	5.36	1.05
Q4	13.83	11.23	11.06	7.41	14.90	12.97	13.90	11.20	1.16
2019	11.40	9.35	10.64	7.14	11.74	9.64	10.04	9.24	1.13
Q1	12.32	10.27	11.79	7.64	13.69	11.25	11.85	10.43	1.24
Q2	12.67	11.14	12.88	8.13	12.34	10.11	10.50	9.81	1.14
Q3	10.84	8.91	10.14	6.80	11.14	9.14	9.44	9.01	1.06
Q4	9.86	7.27	8.00	6.04	9.99	8.19	8.54	7.84	1.08

Source Calculations by research team

the US dollar in 2018 will continue to ease pressure on China's capital outflow. It is expected that the scale of foreign exchange reserves will be expanded to US$3.33 trillion in 2018; it will remain at a scale of US$3.34 trillion in 2019 (Table 2.1).

(b) Growth Rate of Investment in Fixed Assets

In the next two years, the growth rate of fixed asset investment of China will continue to be decelerated owing to the following factors: First, strict financial supervision, enhanced financial deleveraging will inhibit the expansion of loan; Second, higher corporate debt burden and the upward trend in interest rates will inhibit the growth of new investment by enterprises. Third, strict control over the debt risk of local governments will greatly curb the growth of infrastructure investment by local governments. Fourth, investment and financing supervision of the real estate market in the next two years will continue; finally, the growth rate of private investment cannot be quickly rebounded. The fixed assets investment (excluding rural households) is forecasted to increase by 6.57% in nominal terms in 2018, a decrease of 0.63 percentage points over the previous year, and further decline to 5.71% in 2019. It shows that the deceleration of investment in the next two years will restrain the economic growth (Fig. 2.6).

From the sources of funds for investment, the total investment is expected to increase by 6.64% in 2018, an increase of 1.84 percentage points over the previous year; the growth rate of investment in 2019 may fall back to 4.80%. In terms of its

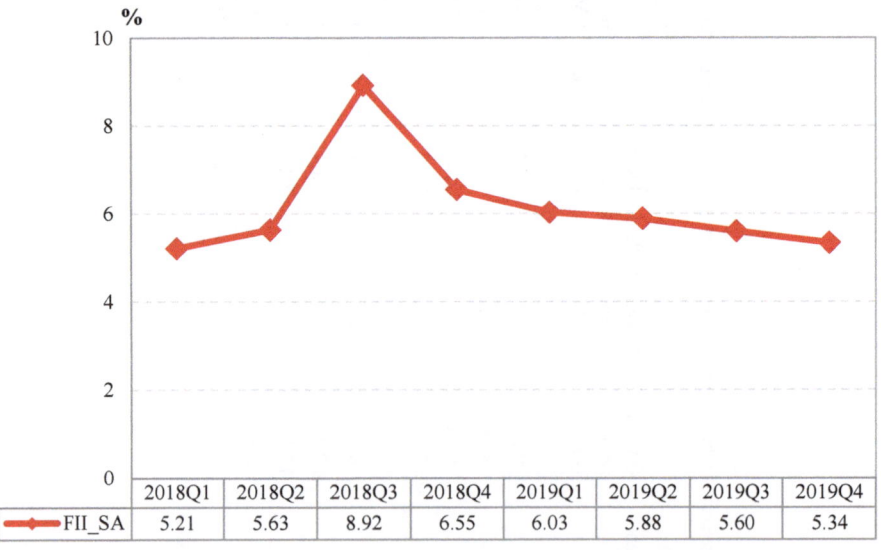

	2018Q1	2018Q2	2018Q3	2018Q4	2019Q1	2019Q2	2019Q3	2019Q4
FII_SA	5.21	5.63	8.92	6.55	6.03	5.88	5.60	5.34

Note: FII_SA indicates seasonally adjusted investment in fixed assets (excluding farmers)

Fig. 2.6 Growth forecast of fixed assets investment (quarterly YoY growth rate). *Source* Calculations by research team. Note: FII_SA indicates seasonally adjusted investment in fixed assets (excluding farmers)

Table 2.2 Forecast of growth rate of fixed assets in the whole society from 2018 to 2019: %

Time	Total investment	Domestic loans	Self-raised funds	Other funds
2018	6.64	5.76	5.02	11.10
Q1	9.27	11.19	8.97	5.83
Q2	6.80	1.21	5.91	10.10
Q3	6.83	5.40	4.12	18.41
Q4	4.06	5.79	1.87	10.50
2019	4.80	4.66	3.16	9.80
Q1	5.81	5.05	4.54	10.31
Q2	4.77	4.64	3.09	10.03
Q3	4.59	4.50	2.92	9.59
Q4	4.13	4.48	2.27	9.30

Source Calculations by research team

composition, investment from domestic loans is expected to increase by 5.76%, which is 2.84 percentage points lower than that in 2017, and it will drop to 4.66% in 2019. The improvement in corporate profit growth will accelerate self-financing investment growth. In 2018, the investment from enterprise self-financing is expected to increase by 5.02%, an increase of 2.72 percentage points over the previous year. This largely avoids a significant drop in investment growth in 2018; the investment growth rate may decline to 3.16% in 2019. The increase in financial supervision will continue to curb investment growth from other funds: this part of investment is expected to increase by 11.10% in 2018, a decline of 0.50 percentage points over the previous year and is further expected to decrease to 9.80% in 2019 (Table 2.2). In general, on the one hand, due to the sustained high growth rate of profit of industrial enterprises, the profitability of enterprises has significantly improved. On the other hand, the growth rate of self-financing by enterprises may gradually recover, while the growth of investment from domestic loans and other components will continue to decline, owing to the continued deepening of financial deleveraging which may inhibit the expansion of shadow banking business and off-balance-sheet operations of financial institutions, and owing to the change in the structure of funding sources for fixed asset investment.

(c) **Resident's Income and Consumption Growth Rate**

In the next two years, the slowdown in the growth rate of labor productivity will continue to curb the growth of residents' real income. The research team predicts that the per capita real disposable income of urban residents is expected to increase by 5.34% in 2018, which is 1.16 percentage points lower than that in 2017, and this income growth rate may recover slightly to 5.64% in 2019. On the other hand, the real cash income per capita of rural residents in 2018 is expected to increase by 8.49%, a decrease of 1.47 percentage points over the previous year, and the income growth rate may slightly rise to 8.85% in 2019 (Fig. 2.7).

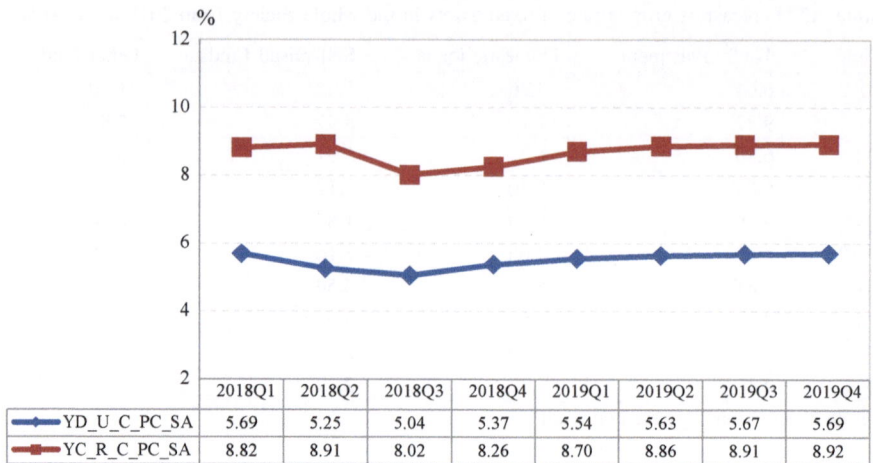

	2018Q1	2018Q2	2018Q3	2018Q4	2019Q1	2019Q2	2019Q3	2019Q4
YD_U_C_PC_SA	5.69	5.25	5.04	5.37	5.54	5.63	5.67	5.69
YC_R_C_PC_SA	8.82	8.91	8.02	8.26	8.70	8.86	8.91	8.92

Note: YD_U_C_PC_SA represents the seasonally adjusted per capita disposable income of urban residents

(constant price), and YC_R_C_PC_SA represents the seasonally adjusted per capita cash income

of rural residents (current price)

Fig. 2.7 Forecast of urban and rural residents' income. *Source* Calculations by research team. Note: YD_U_C_PC_SA represents the seasonally adjusted per capita disposable income of urban residents (constant price), and YC_R_C_PC_SA represents the seasonally adjusted per capita cash income of rural residents (current price)

Affected by the increase in residents' income, the total retail sales of nominal social consumer goods is expected to increase by 10.60% in 2018, a slight increase of 0.30 percentage points over the previous year; with price adjustment, real household consumption will increase by 6.79%, a decrease of 0.44 percentage points over the previous year. In 2019, the nominal growth rate of total retail sales of social consumer goods will drop to 10.39%, and the actual growth rate of household consumption will rise to 7.09%. From a quarterly perspective, both the nominal growth rate of total retail sales of social consumer goods and the growth rate of actual household consumption have shown a trend of "first decline and then increase" (Fig. 2.8).

In summary, the forecast based on the CQMM model shows that the sustained growth of the world economy will continue to stimulate foreign trade growth of China in the next two years, but the slowdown in domestic investment growth will continue to inhibit economic growth. On the whole, the economy is expected to maintain a growth level of 6.50% or more, and the trend of "steady and slowing down" in economic growth will continue.

(a) It is expected that real GDP will increase by 6.73% in 2018, a decrease of 0.17 percentage points over the previous year; CPI will increase by 2.13%, an increase of 0.53 percentage points over the previous year; PPI will increase by 4.64%, a decrease of 1.66 percentage points over the previous year. The price

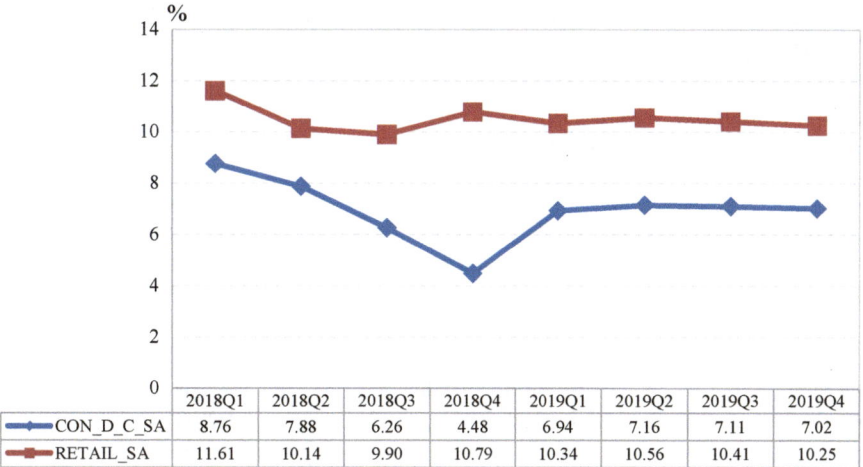

	2018Q1	2018Q2	2018Q3	2018Q4	2019Q1	2019Q2	2019Q3	2019Q4
CON_D_C_SA	8.76	7.88	6.26	4.48	6.94	7.16	7.11	7.02
RETAIL_SA	11.61	10.14	9.90	10.79	10.34	10.56	10.41	10.25

Note: CON_D_C_SA represents the seasonally adjusted total growth in consumer spending (constant price);

RETAIL_SA represents the seasonally adjusted total retail sales of consumer goods (current price)

Fig. 2.8 Forecast of year-on-year growth rate of consumption. *Source* Calculations by research team. Note: CON_D_C_SA represents the seasonally adjusted total growth in consumer spending (constant price); RETAIL_SA represents the seasonally adjusted total retail sales of consumer goods (current price)

level remains in a controllable range, there is no obvious inflation risk, and the disparity between the growth rate of PPI and CPI will also be reduced.

(b) The supervision over financial markets to prevent and control financial system risks and the control of local government debt risks, together with the difficulty of rapid rebound of private investment, will continue to curb investment growth in the next two years. Based on current prices, investment in fixed assets (excluding rural households) is expected to increase by 6.57% in 2018, a decrease of 0.63 percentage points over the previous year.

(c) The growth rate of actual disposable income of urban and rural residents may decline slightly, resulting in a decrease in the growth rate of actual household consumption. In 2018, the actual growth rate of resident consumption is expected to be about 6.79%, a decrease of 0.44 percentage points from 2017. In 2018, the total retail sales of nominal social consumer goods are expected to increase by 10.60%, an increase of 0.30 percentage points over the previous year.

(d) The recovery of the world economy will continue to drive the steady growth of China's foreign trade. In 2018, total export (current US dollar value) is expected to increase by 9.65%, an increase of 1.75 percentage points over the previous year; and the total value of imports (current US dollar value) will increase by 12.32%, a decrease of 3.58 percentage points from 2017. Stabilization of the RMB against the US dollar will continue to ease pressure on China's capital outflow. It is expected that the scale of foreign exchange reserves will expand to US$3.33 trillion in 2018.

2.3 Analysis of Potential Factors Affecting Forecast Results

2.3.1 Forecast Errors of Several Indicators of China's Economy in 2017

Comparing the differences between the forecasted and actual values in 2017, the research team finds that the forecasted value of actual GDP growth is underestimated by 0.1%. The reasons for this underestimation are as follows: First, the model overestimates the price level; the forecasted value of the CPI growth rate is overestimated by 0.12 percentage points from the actual value. Second, the model underestimates the growth of foreign trade. In 2017, the forecasted values of the growth rate of exports and imports (the constant value of RMB) were underestimated by 0.92 and 0.51 percentage points respectively from the actual value. Third, the model overestimates the nominal growth rate of investment in fixed assets that does not include rural households, and its forecast value is about 1.25 percentage points higher than the actual value (Table 2.3).

2.3.2 Potential Factors Affecting the Outcome of This Forecast

First, if the economic growth in the United States and the Eurozone in the next two years are lower than the growth rate assumed by the research team, this forecast may overestimate the growth of export, and then overestimate the economic growth.

Table 2.3 Differences between forecasted values and actual values of some indicators in fall 2017

Indicator	Forecast value	Actual value	Forecast value—actual value
GDP	6.8	6.9	−0.1
CPI	1.72	1.6	+0.12
PPI	6.24	6.3	−0.06
RMB-denominated export growth rate	9.88	10.8	−0.92
RMB-denominated import growth rate	18.19	18.7	−0.51
Investment in fixed assets (excluding farmers)	8.55	7.3	+1.25
Current total social retail sales	10.69	10.2	+0.49
Urban residents' actual disposable income growth rate	6.39	6.5	−0.11

Note The disparity indicates the forecasted value minus the observed value

Second, the research team assumes that the growth rate of M2 in the four quarters of 2018 will be "low first and then high", and the annual growth rate will be about 8.5%. However, the data released by IMF in January 2018 indicated that the growth rate of M2 in January was approximately 8.6%, exceeding the research team's setting for M2 growth in the first quarter. Considering that the beginning of a year is often the fastest time for money growth, the fall in M2 growth in February and March of 2018 should be a big probability event. Therefore, the research team no longer modifies the assumption about M2 growth. However, if the monetary authorities change the strict financial supervision in the future and accelerate the deleveraging policy, which may lead to a rapid recovery of M2 growth, then the model may underestimate the economic growth. In addition, abnormal changes in the growth rate of fiscal expenditure may also lead to greater errors in the estimates.

Third, if the real appreciation/depreciation of the RMB against the US dollar is too high, it will lead to model overestimation/underestimate the growth rate of exports and imports, leading to overestimation/underestimation of economic growth forecast values.

Chapter 3
Policy Effect Simulation

3.1 Background Analysis

In recent years, the leverage ratio of economy, finance, and non-financial sectors hasrisen rapidly and is at high levels, which have led to an increase in the possibility of systemic risks in the economy. Among them, due to the existence of "soft budget constraints" and implicit guarantees from government departments, the leverage ratio of state-owned sectors is ease to rise and hard to fall, creating a ratchet effect.[1] Although at the end of 2015, the Central Economic Working Conference had already proposed to reduce leverage ratio, but in the reality, the deleveraging process was not easy: First, from a macro perspective, the leverage ratio of non-financial enterprises has declined recently, but still at a relatively high level (Fig. 3.1); Secondly, in non-financial state-owned enterprises, the leverage ratio of state-owned industrial enterprises has recently declined, but the leverage ratio of all state-owned enterprises is still at a high level (Fig. 3.2). In the end, the progress of "deleveraging" in the financial sector is still not optimistic. In the past years, the central bank strictly controlled the growth of money supply, tightened the supply of funds, and reduced the financial leverage ratio; on the other hand, it strengthened financial supervision and strictly controlled the rapid expansion of internal credit derivatives in the financial sector, especially the expansion of loan of banks and non-banking institutions, to prevent investment from being diverted out of the real economy. It is difficult to reverse the situation in which a large amount of capital is idling within the financial system without entering the real economy. However, although the scale of the new RMB loans in the past two years hasbeen expanding year by year, the proportion of loans to the real economy is still less than half.

---

[1]Research team of Center for Macroeconomic Research Center, Xiamen University: "China's Macroeconomic Forecasting and Analysis—Fall 2017 Report."

© Springer Nature Singapore Pte Ltd. 2018
Center for Macroeconomic Research at Xiamen University, *China's Macroeconomic Outlook*, Current Chinese Economic Report Series,
https://doi.org/10.1007/978-981-13-1005-8_3

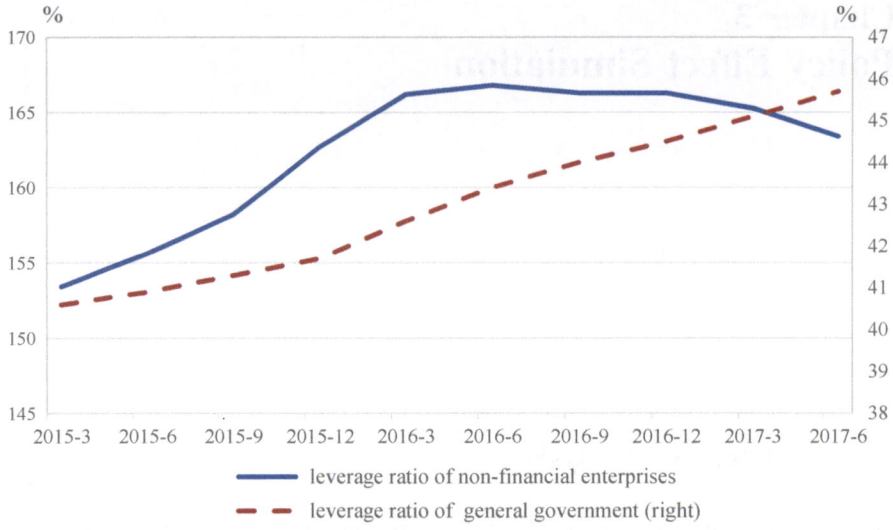

Fig. 3.1 Leverage ratio of non-financial enterprises and general government (as of June 2017). *Source* BIS

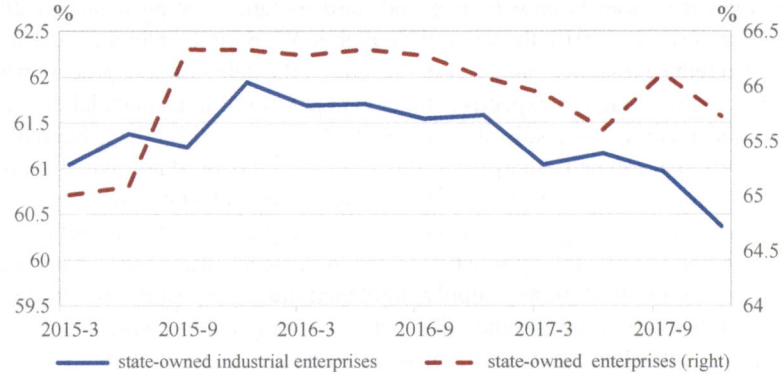

Fig. 3.2 Changes in asset-liability ratio of state-owned enterprises. *Source* CEIC

In 2018, the financial and non-financial sectors to reduce leverage ratio, prevent and resolve major risks are still the core concerns of various macroeconomic policies.[2] At the 2018 Economic Working Conference held on December 8, 2017 by the Political Bureau of the CPC Central Committee, it was clearly pointed out

---

[2]At the 2018 annual meeting of the World Economic Forum on January 24, 2018, Liu He, member of the Political Bureau of the CPC Central Committee and director of the Central Leading Group of Finance and Economics, said that in the face of the current outstanding financial risks, China will strive for the next three years or so. The effective reducing of the macro leverage ratio, improving the adaptability of the financial structure, and strengthening the economic capabilities of

that the prevention of major risks should effectively control the macro leverage ratio, the economic capabilities of the financial service entities should be strengthened, and the risk prevention work should achieve positive results. In addition, the State Council executive meeting held on February 7, 2018 also pointed out that it was necessary to continue to regard state-owned enterprises as a top priority, and in combination with measures such as state-owned enterprise reforms, reduction of production overcapacity, reduction of cost and leverage ratio. These intensive and targeted statements underscore the policy authorities' determination to reduce leverage ratio. It can be said that in the next few years, reducing the leverage ratio has become imperative.

Since 2015, the sluggish growth in private investment in competitive sectors had prompted local governments, state-owned and state-controlled enterprises to increase their investment to stabilize investment growth. By 2017, the proportion of state-owned and state-controlled enterprise investment in total investment had increased significantly to 36.9%, and it had gradually become an important force for steady economic growth (Fig. 3.3). Because the state-owned enterprises had long had a "natural advantage" in obtaining financial resources, and there were "soft budget constraints" and implicit guarantees from government, the leverage ratio of state-owned enterprises had risen easily. In this context, given the policy basis for the reduction of leverage ratio of state-owned enterprises, a natural concern is: reduction of leverage ratio is likely to lead to a shrink in the growth of state-owned investment, and to what extent will it affect economic growth?

In general, the process of deleveraging is accompanied by a slowdown in loan growth and shrinking aggregate demand[3]; Ma Yong et al. (2016) also showed that deleveraging has a significant negative effect on economic growth.[4] In other words, to prevent the systemic financial risks, it is necessary to effectively control macro-leverage ratio, but this will inevitably result in a decline in the growth rate of state-owned investment and shake the important pillars of steady growth; especially in the context of the rapid rebound of private investment growth. The impact on economic growth may be particularly evident.

Consequently, the research team used the CQMM model to simulate the impact of the deleveraging policy of state-owned enterprises in the next two years. The key point is to simulate the impact of deleveraging of state-owned enterprises, which may affect the growth of investment and economy, thus providing a quantitative analysis basis for the macro effect of the deleveraging policy, and assessing the profit and loss of the deleveraging policy with policy suggestions.

---

the financial services entity will allow systemic risks to be effectively prevented and the economic systems will enter virtuous circle.

[3]McKinsey Research Report (2010): *The looming deleveraging challenge*, https://www.mckinsey.com/global-themes/employment-and-growth/the-looming-deleveraging-challenge.

[4]Ma Yong, Tian Tuo, Zhuo Zhuoyang, Zhu Junjun (2016): "Financial Leverage, Economic Growth and Financial Stability", "Financial Studies", 06(6), pp. 37–51.

Fig. 3.3 Trends in the growth rate of state-owned investment and private investment (accumulated year-on-year). *Source* CEIC

3.2 Simulation Scenario Setting

The research team designed two types of simulation scenarios.

Scenario 1: Assuming a substantial advancement of the deleveraging policy in the next two years, the leverage ratio of state-owned industrial enterprises will fall to 56% quarter-over-quarter within eight quarters, basically returning to the level before the outbreak of the financial crisis in 2007 (Fig. 3.4). The setting of this simulation scenario can assess impact of deleveraging of the state-owned enterprises on investment growth and economic growth.

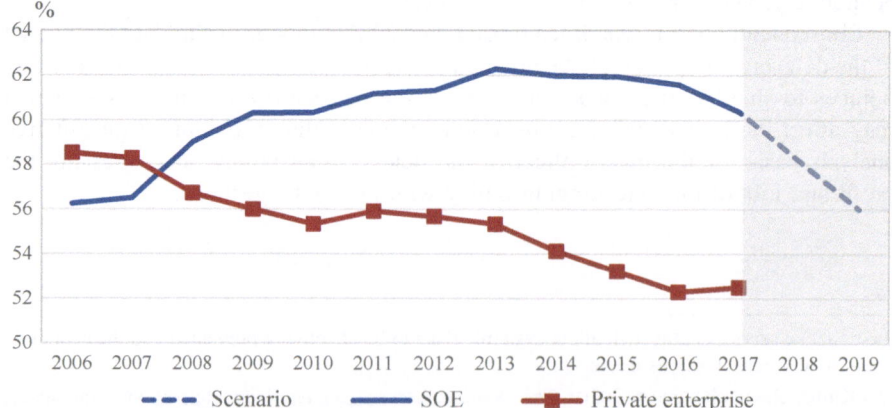

Fig. 3.4 Scenario 1: reduce the asset-liability ratio of state-owned enterprises at a uniform rate. *Source* WIND, CEIC

Scenario 2: In order to ensure that the economic growth rate is stable at a level of around 6.5%, then how much should private investment increase to fully absorb the negative impact of the deleveraging of state-owned enterprises on economic growth? The setting of this simulation scenario can assess the "compensation effect" of private investment on deleveraging policies, thus providing support for policy analysts to formulate relevant policies to promote the recovery of private investment growth.

3.3 Policy Simulation Results

3.3.1 Scenario 1: Macroeconomic Impact of Leverage Reduction in State-Owned Enterprises

The simulation based on the CQMM model found that when other conditions remain unchanged:

First, the de-leveraging of state-owned enterprises will drive down the investment growth of state-owned enterprises. As the growth of state-owned enterprises' investment has largely been the result of high leverage ratio, when leverage ratio and the loan scale are reduced, the investment of state-owned enterprises will shrink accordingly. In the simulation scenario, the growth rate of state-owned investment in 2018 will drop to −1.04%, which is a substantial drop of 6.34 percentage points from the baseline forecast. This growth rate may further decline to −2.89% in 2019, which is a decrease of 10.19 percentage from the forecasted value (Fig. 3.5).

Second, when other conditions remain the same, the shrinking of state-owned investment growth will significantly reduce the growth rate of investment in fixed assets across the society. In 2018, the growth rate of fixed asset investment will fall to 5.09%, a decrease of 1.48 percentage points from the forecasted value; in 2019, this growth rate further will fall to 2.08%, which is 3.63 percentage lower than the forecasted value (Fig. 3.6).

Finally, when other conditions remain unchanged, the decline in fixed asset investment growth will significantly reduce the growth rate of GDP. The GDP growth rate in 2018 will likely drop to 5.92%, which is 0.81 percentage points lower than the forecasted value; it will further decline to 5.37% in 2019, 1.23 percentage points lower than the forecasted value, and will be lower than the 6% level in two years (Fig. 3.7).

The above simulation results show that under certain conditions, deleveraging will a negative impact on economic growth. In order to maintain growth, state-owned enterprises have accelerated investment growth through high leverage ratio. The reduction of the leverage ratio will inhibit the growth of state-owned enterprises' investment. With the rapid growth of non-government investment, it will be difficult to quickly increase the growth of investment in the entire society, and the economic growth rate will slow down. This will result in a dilemma:High

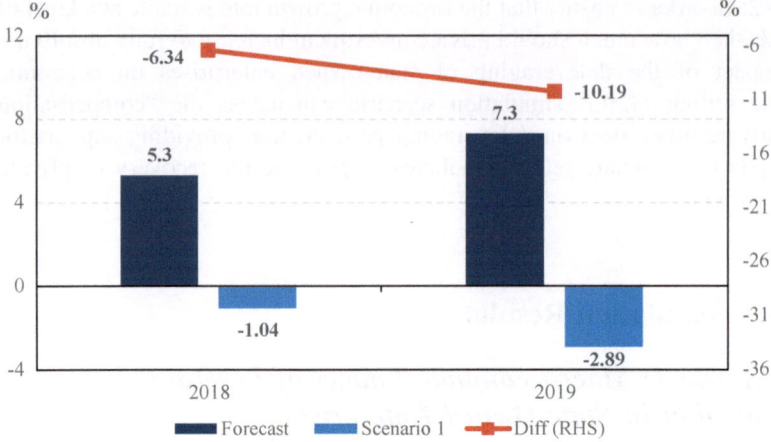

Fig. 3.5 Changes in state-owned investment growth in the simulation scenario. *Source* Calculations by the research team. Note: Forecast represents the baseline forecast value; Scenario1 represents the forecasted value of the simulated scenario; Diff is the difference between the forecasted value of scenario 1 and the baseline forecasted value

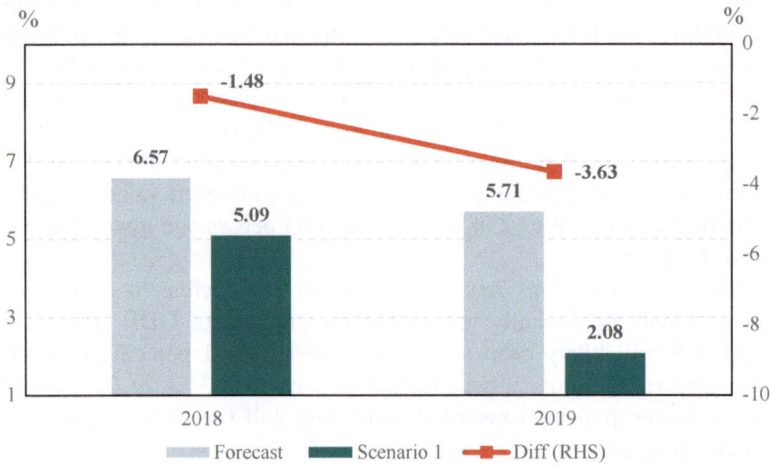

Fig. 3.6 Changes in growth rate of investment in fixed assets. *Source* Calculations by the research team. Note: Forecast represents the baseline forecast value; Scenario1 represents the forecasted value of the simulated scenario; and Diff is the difference between the scenario 1 forecasted value and the baseline forecasted value

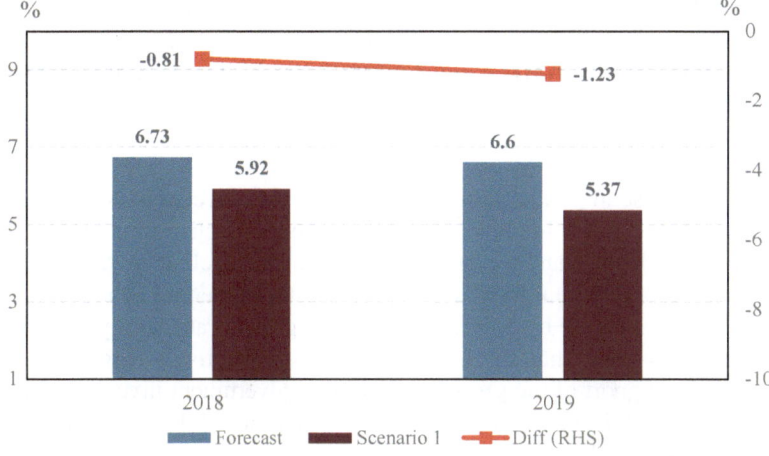

Fig. 3.7 Changes in GDP growth in the simulation scenario. *Source* Calculations by the research team. Note: Forecast represents the baseline forecast value; Scenario1 represents the forecasted value of the simulated scenario; and Diff is the difference between the scenario 1 forecasted value and the baseline forecasted value

leverage ratio of state-owned enterprises may increase the fragility of the financial system and induce economic systemic risks, and the deleveraging may threaten the stability of economic growth. However, it is not impossible to reduce leverage ratio and to maintain growth simultaneously. The key is whether the macroeconomic policies can effectively stimulate the growth of private investment. Under the circumstances that the world economy will recover steadily in the next two years, if the domestic business environment and policies and measures are adjusted to stimulate the growth of private investment, then the growth of private investment will probably offset the shrinking of state-owned enterprises due to deleveraging, and ensure steady economic growth.

Scenario 2: Encouraging the private investment and achieving the goals of deleveraging and steady growth

Under this scenario, in order to stabilize the economic grow that the level of 6.5%, by how much should the growth of private investment be raised, such that the economy can fully absorb the negative impact of deleveraging of state-owned enterprises on economic growth.

The simulation results show that, first of all, it is necessary to reduce the leverage ratio of state-owned enterprises and stabilize the economic growth rate at 6.5% in the next two years. Then, in 2018, the growth rate of private investment needs to be raised to 13.35%, which is 2.8 percentage points higher than the forecasted value; the growth rate of private investment in 2019 needs to be maintained at 14.42%, which is 4.6 percentage points higher than the forecasted value (Fig. 3.8). This means that if there is no corresponding increase in the growth rate of private investment, then the de-leveraging of state-owned enterprises may lower the economic growth. Conversely, while promoting the reduction of leverage ratio of state-owned enterprises, effectively stimulating the growth of private investment, the goal of stabilizing economic growth can be achieved. This simulation result provides support for empirical analysis by measuring the "compensation effect" of private investment.

Secondly, the increase in private investment growth will effectively compensate for the state-owned enterprises' decline in leverage and the decline in state-owned investment, which will, to a large extent, slow down the decline in fixed asset investment growth. In 2018, total fixed asset investment in the entire economy will increase by 6.13% and increase by 5.35% in 2019. The growth rate is only 0.44 and 0.36 percentage points lower than the forecasted value, respectively (Fig. 3.9).

Finally, under the support of the growth rate of non-government investment, the GDP growth rate in 2018 is expected to reach 6.53%, and it will be 6.5% in 2019, which is only 0.2 and 0.1 percentage points lower than the forecasted value respectively (Fig. 3.10).

In summary, while promoting the deleveraging policy of state-owned enterprises, the goal of stabilizing economic growth can be achieved at the same time, as long as it can effectively stimulate the growth of private investment. In one word, reducing leverage ratio and stabilizing growth are not mutually exclusive.

In fact, the steady economic growth is also very important for promoting the deleveraging policy smoothly. The macro leverage ratio is usually measured by the

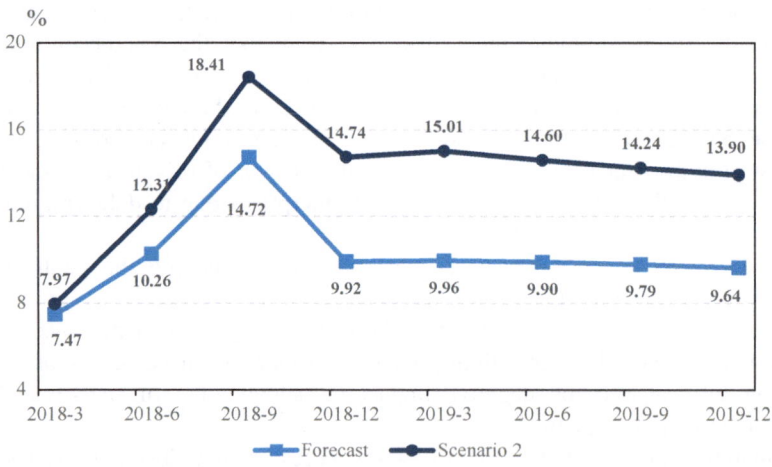

Fig. 3.8 Growth of private investment (simultaneous year) under simulated scenario 2. *Source* Calculations by the research team. Note: Forecast represents the baseline forecast value; Scenario2 represents the forecast value under simulation scenario 2; Diff is the difference between the scenario II forecast value and the baseline forecast value

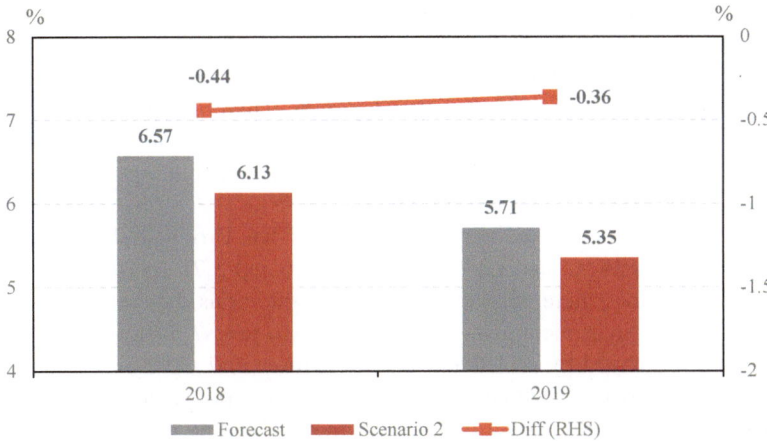

Fig. 3.9 Changes in growth rate of investment in fixed assets under scenario 2. *Source* Calculations by the research team. Note: Forecast represents the baseline forecast value; Scenario2 represents the forecast value under simulation scenario 2; Diff is the difference between the scenario II forecast value and the baseline forecast value

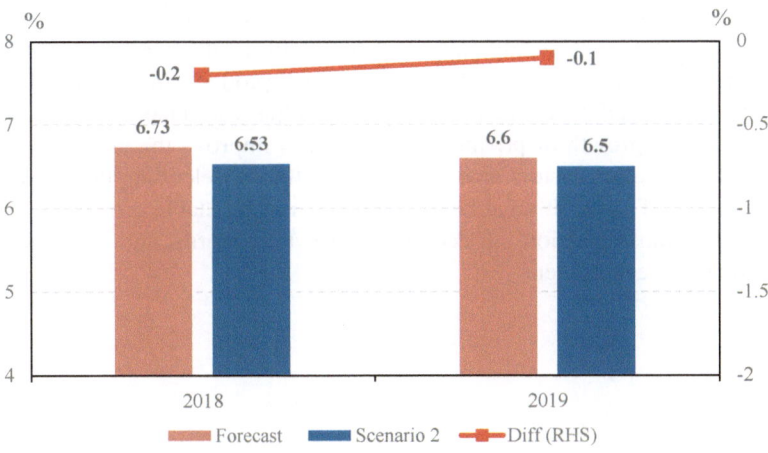

Fig. 3.10 Changes in GDP growth under simulated scenario 2. *Source* Calculations by the research team. Note: Forecast represents the baseline forecast value; Scenario2 represents the forecast value under simulation scenario 2; Diff is the difference between the scenario II forecast value and the baseline forecast value

index of debt divided by GDP. The denominator of the index actually reflects the tolerance of debt guaranteed by growth prospects. This provides two ways to reduce leverage: one is to reduce the numerator by reducing debt; the other is to increase the denominator by promoting economic development. Therefore, while delever-aging, it is undoubtedly an active measure to reduce the numerator and increase the denominator by promoting private investment and promoting economic development.

Since 2015, the growth rate of private investment has dropped sharply and remained at a low level (Figs. 3.1, 3.2, 3.3, 3.4, 3.5, 3.6 and 3.7). In 2017, it only increased by 6%. So, does there exist environmental conditions that can drive the rapid growth of private investment growth in the next two years? First of all, in the foreign trade, the role played by private enterprises is more and more important. In 2017, the private enterprises accounted for 44.4% of total exports. In the next two years, the growth of the world economy will inevitably stimulate the expansion of the export of private enterprises, which will in turn motivate the growth of their investment. Second, the financial resources released by the deleveraging of state-owned enterprises and the measures of adjusting the structure of financial resources to serve the real economy will, to a large extent, expand the private investment. Third, further deepening institutional reforms will reduce the long-standing investment barriers against private enterprises in the past, thereby expanding the investment areas of private enterprises; Finally, the renovation, cultivation and improvement of the business environment will also create favorable conditions for the expansion of private investment from this year.

Under the background of supply-side structural reform, the rapid growth of private investment can effectively adjust the industry and product structure to adapt to changes in the demand structure that constantly adjusts as income increases. More importantly, the growth of private investment can improve the efficiency of the allocation of production factors among industries, thus accelerating the growth of labor productivity. The result must be that while achieving stable growth, it can also promote the transformation of economic growth patterns and ultimately achieve high-quality development.

Chapter 4
Policy Recommendations

In 2017, China's economy continued to grow steadily, with a growth rate of 6.9%. The rapid growth of infrastructure investment and the rapid rebound in the growth of imports and exports were two of the most important factors in stabilizing economic growth. On the one hand, the economic structure continued to be optimized, the share of the tertiary industry continued to increase, and the share of high-end manufacturing investment continued to rise. On the other hand, the leverage ratio of the financial and non-financial sectors has risen rapidly and hasbeen at a high level, which has greatly increased the possibility of systemic risks arising from the economy. This not only indicated that the factors behind the economic growth were more complex, but also indicated that the prospects for economic growth may be even more uncertain.

To some extent, the growth rate of non-government investment since 2015 had been difficult to rapidly rebound. This was the main reason why macroeconomic problems have not been fundamentally resolved. In 2017, the profit growth of state-owned and state-controlled (especially large and medium-sized) industrial enterprises involved in the upstream production materials sector and monopoly industries began to increase rapidly due to barriers to entry of private enterprises, while the profit growth of private enterprises in the competitive sector increased relatively mildly. It shows that the trend of rapid rebound of non-government investment will be hard to continue. The policy authorities is required to work hard to resolve some potential problems.

First of all, it was difficult for the growth of private investment to quickly rebound. Therefore, local governments at all levels had to increase their investment in infrastructure; at the same time, the financial sector had also to expand loan supply to state-owned enterprises. On the one hand, due to the existence of "government implicit guarantees" and "soft budget constraints" of the fiscal (state-owned enterprises), the funds released by the loose monetary policies strongly supported the local financing platforms and state-owned zombie enterprises with low efficiency through the shadow banking system. This in turn pushed up the leverage ratio of the government and state-owned enterprises. On the other

© Springer Nature Singapore Pte Ltd. 2018
Center for Macroeconomic Research at Xiamen University, *China's Macroeconomic Outlook*, Current Chinese Economic Report Series,
https://doi.org/10.1007/978-981-13-1005-8_4

hand, the shrinking profit growth of the industrial sector directly restrained the growth of fiscal revenues of local governments. Therefore, local governments were more relying on revenue from land sales, driving up the price of real estate market, further promoting the flow of financial resources to real estate market, and resulting in investment from being diverted out of the real economy. It was difficult to reverse the situation in which a large amount of capital was idling within the financial system without entering the real economy. In addition, the long-term tilt of the financial policy orientation toward state-owned enterprises, which not only rapidly increased the debt of non-financial state-owned enterprises but also inevitably "crowded out" the financial resources available for private investment, further inhibited the rebound of private investment.

Second, the decline in private investment growth was directly reflected by the shrinking growth in manufacturing investment. This not only inhibited the expansion of industrial production, but also was not conducive to the upgrading of industrial structure and industrial efficiency. In 2017, the share of manufacturing investment in total investment continued to decline to 30.7%, which was 3.6 percentage points lower than the highest level in 2012. At the same time, the contribution rate of the industrial sector to GDP growth declined sharply: in 2017, the contribution rate of secondary industry to GDP growth fell sharply to 36.2% in 2017, which was more than 50% averagely before the financial crisis. It showed that industrial production efficiency needs to be rapidly improved, and the industrial structure needs to be quickly adjusted and upgraded. However, the long-term sustained downturn in private investment growth would inevitably hinder the intensive expansion of the manufacturing industry and its transformation and upgrading.

Third, it was difficult for the growth of non-government investment to rebound. This not only inhibited the growth of investment in fixed assets, but also led to further imbalances in investment structure and further reduces investment efficiency. Since 2015, with the deceleration of private investment, the changes of the ownership structure showed that the share of investment by state-owned enterprises had continued to rise and the share of private investment had continued to decline. In terms of industry structure, the share of investment in manufacturing had continued to decline and the share of infrastructure investment had decreased, and the proportion of real estate investment remained high. This imbalanced investment structure undermined the economic efficiency of investment and increased the fragility of the financial system.

Fourth, although the residents' income grew steadily in 2017, the decline in labor productivity growth in recent years had been inhibiting the rapid growth of the residents' real income. In the foreseeable future, the rapid growth of labor productivity in China's industry, especially the manufacturing industry, will inevitably be an important guarantee for labor productivity and the rapid increase in the real income growth of residents. To reverse the decline in the growth rate of labor productivity, China must rely on the rapid increase in labor productivity in the manufacturing industry, and effective investment expansion (especially private investment) is the necessary means to promote the transformation and upgrading of the manufacturing industry and thus accelerate labor productivity growth. Therefore, the current risk of continued decline in the growth rate of private investment is significant, especially the decline in the growth rate of investment in manufacturing should be given great attention.

In 2018, in the financial and non-financial sectors, to reduce leverage ratio, and to prevent and resolve major risks are still the core concerns of macroeconomic policies. First of all, the regulatory oversight of local finances will continue to be strengthened, and the loopholes through which local governments could borrow money abnormally will have basically been blocked. As a result, the rapid expansion of infrastructure investment will be unsustainable. Second, the regulatory authorities will start to rectify all kinds of financial chaos. Especially, all kinds of violations related to real estate development loans will be severely punished, which will largely inhibit the growth of real estate investment. Finally, combined with the reform of state-owned enterprises, the reduction of production overcapacity and the reduction of cost, the process of deleveraging of state-owned enterprises will also be substantially advanced. These measures to prevent and control financial risks will inevitably increase downward pressure on investment growth. If private investment growth cannot rebound substantially, then the deceleration of investment in fixed assets will inhibit economic growth. With the economic slowdown, the demand for expansion of local government infrastructure and its reliance on the revenue from land sales, the financial sector's preference for state-owned enterprises and the real estate industry may recover. It is difficult to reverse the situation in which a large amount of capital is idling within the financial system without entering the real economy. It is even more difficult for fundamental financial and non-financial state-owned enterprises to eliminate the two financial risks. Therefore, while preventing and controlling financial risks, it is necessary to further investigate how China can effectively stimulate the growth of private investment, improve the efficiency of resource allocation, and thus promote a series of issues such as structural adjustment, transformation and upgrading of manufacturing industry, and accelerating the growth of labor productivity.

In the past years, the research team has focused on the issue of the deceleration of private investment from various aspects, such as: the barriers against private enterprise, and the decline in investment yields caused by the unclear prospects for corporate debt and profit growth. Under the background of preventing and controlling financial risks, some new factors that restrict private investment, especially real economic investment growth, cannot be ignored. First, with high corporate debt burden, the financial sector is more cautious about lending to the real economy in order to prevent risks. Second, the rising cost of bank funds caused by the prevention and control of financial risks will further weaken the financial sector. It will further weaken the willingness of financial institutions to invest in the real economy. At the same time, it will strengthen its willingness to invest in real estate and state-owned enterprises, etc. Third, financial deleveraging has effectively suppressed non-financial institutions to obtain investment funds from other sources, and the channels for private enterprises to obtain funds have also narrowed accordingly. At the same time, it is difficult for the private enterprise's profit growth to rebound rapidly and the growth rate of self-financing enterprises continues to decline. These factors add up to a variety of institutional defects that have existed for a long time, all of which have made it more difficult for private investment to rebound quickly.

On the other hand, according to the predictions on the macroeconomic situation and the policy simulation analysis, we believe that the rapid rebound of non-governmental investment has had many favorable conditions: First, the continued growth of the world economy will inevitably pull in the next two years. The expansion of the export of private enterprises in China will in turn motivate its investment growth. The second is that the financial resources released by the state-owned enterprises because of deleveraging and the adjustment of the structure of financial resources to make them truly serve the real economy will, to a large extent, meet the funding needs for private investment expansion. Third, further deepening the reform will reduce the previous long-term investment barriers to private enterprises, which will expand the investment area of private enterprises. Finally, the promotion of various measures such as the renovation, cultivation, and improvement of the business environment of the country this year will also create favorable conditions for the expansion of private investment. Based on this, policy authorities should make full use of the opportunity of the domestic and foreign economies and comprehensively apply various policies and measures. While preventing and controlling financial risks, the policy should focus on stimulating the private economy and promoting the rapid rebound of private investment. In the long run, only by relying on the market forces and improving the efficiency of the allocation of production factors among different ownership enterprises and different industries can China ensure a rapid increase in labor productivity. While preventing and controlling financial risks, China can stabilize economic growth and achieve high quality of development.

Therefore, it is necessary to rectify the financial governance in the next two years, but at the same time, China should also pay attention to adopting appropriate measures to protect the loan demand for expansion of investment in manufacturing, and take advantage of the opportunities for continuous improvement of the world economy in the next two years to promote China's manufacturing industry. The intensive expansion and its transformation and upgrading will consolidate the industrial base of economic growth, and will accelerate the growth of labor productivity, and ultimately achieve a long-term and rapid increase in the real income of residents.

The policy suggestions are as follow.

(a) Stable growth basically requires the rapid growth of effective investment and the improvement of investment efficiency. While the current economic growth momentum has not been completely transformed yet, the pressure of continued declining investment growth should be paid attentions. Therefore, while preventing and controlling financial risks, China must start with adjusting the structure of financial resources to increase the share of newly added RMB loans from non-financial enterprises and organizations. When the economic growth rate stabilizes, it is necessary to fully satisfy the expanded financial resources. When economic growth stabilizes, it is necessary to make the expanded financial resources fully meet the needs of the real economy, especially the expansion of private investment demand, so that finance can really serve the real economy, and improve investment efficiency by increasing the proportion of private investment.

(b) The rapid increase in labor productivity requires the intensive expansion of the manufacturing industry and its transformation and upgrading. Its driving force will certainly come from private investment. The current over-concentration of financial resources in the real estate industry and the preference for state-owned enterprises has made it difficult for private manufacturing enterprises to obtain financial support. The distortion of the financial structure has resulted in the loss of efficiency. If private investment growth continues to be sluggish, it will inevitably impede the structural upgrading of the manufacturing industry and curb the increase in labor productivity. Therefore, while preventing and controlling financial risks, it is necessary to speed up the market-oriented reform of the capital market and improve the efficiency of allocating financial resources among industries so as to promote the adjustment of investment structure and effectively meet the demand for private investment.

(c) Adhere to the basic tone of "stabilizing and tightening monetary policy". While controlling the total amount of money supply, China will achieve stable economic growth through the structural adjustment of financial resources. The forecasting results of the research team show that the growth rate of the two types of M2 is maintained at the level of 8.2–8.5%, and the growth rate of GDP can be maintained at a growth rate of over 6.5%. Therefore, monetary policy should remain "steady and tight", strictly control the money supply, and actively coordinate macro-prudential supervision and financial stability policies to achieve effective prevention and control of financial risks. In addition, it is necessary to reform the financial supervision system, break the rigid payment, give full play to the capital market's role in reducing leverage ratio and optimize the allocation of resources, strengthen the supervision of commercial banks and prevent them from using non-bank financial institutions to provide loans for industries with production overcapacity. Market-oriented, legalized debt-to-equity transformation, in order to achieve the purpose of deleveraging, continue to strengthen the supervision of shadow banking, combined with macro-prudential assessment system to strictly control the expansion of shadow banking and reduce unregulated financing behavior, and actively guide shadow banking to give full play to the function of serving the real economy.

(d) While reducing the debt ratio of non-financial state-owned enterprises, China should speed up the establishment of a modern fiscal and taxation system and form a hard constraint on the budget of local governments (state-owned enterprises). The construction of a hard budget constraint for local governments (state-owned enterprises) is an important institutional construction that comprehensively deepens institutional reforms and eliminates systemic financial risks. To resolve the soft budget constraints of local governments and to control their ability to expand debt, China must establish compatible mechanism between the fiscal revenue and expenditure of a local government, and a standardize their bond issuance system.

(e) China must change the old development pattern which over-stresses on economic growth. The government should gradually adjust its function and shift it from economic growth to public service and public-government governance so that the market can do market affairs and let the government do government affairs.

(f) By improving laws and regulations, and ensuring full protection of state-owned assets, China will privatize some state-owned enterprises in the competitive sector by auction. At the same time, China will steadily promote the reform of the mixed ownership system of state-owned enterprises to promote the competitive field of state-owned monopoly industries, and to stimulate private investment.

(g) To stimulate the private investment, it is necessary to boost private entrepreneurs' confidence in the domestic economy. The fundamental of boosting confidence is to comprehensively deepen reforms, improve the domestic economic environment, improve the business environment, protect entrepreneurial spirit and corporate property rights, and support entrepreneurs to concentrate on innovation and entrepreneurship.

In summary, we believe that effective measures to promote the rapid recovery of non-government investment will help solve some of the problems and help advance the structural reforms on the supply side. In the next stage, "promoting reforms through openness, promoting growth through innovation, improving efficiency through competition, and ensuring employment through demand" should be the preconditions for China's economic development towards high quality. To this end, supply-side structural reforms should be proactively promoted to the market of production factors, and through the market-oriented reforms, the financial capital market should be continuously improved, and the allocation efficiency of financial resources should be improved through the reduction of institutional transaction costs and the promotion of private investment.

Chapter 5
Comments and Discussions

5.1 The Basic Connotation of High Quality Development

Tongsan Wang
Institute of Quantitative and Technical Economics, Chinese Academy of Social Sciences, Beijing, China

The report by the research team of CQMM in Xiamen University emphasized the precision, which was indicated by the index since it was precise to percentile. The prediction of Taiwan's leadership election is also precise to percentile, but how to test it? I suggest that Xiamen University should take this into consideration. Of course, if you are predicting the real Chinese economy and using the real Chinese data, your predication which is accurate to percentile will have the scientificity, but how to achieve this in a real test? This is my suggestion.

(1) The Economic Growth Trend in China

Why can we achieve a pretty good development in 2017? A very significant reason is that the contribution of external demand was negative in 2016 and it becomes positive this year. We should pay great attention to the impact of external environment on Chinese economy. Now China had entered a new normal, and the interaction relationship and effect between the Chinese and the World's economy becomes more and more obvious. When studying the world economic environment, there is a question we should pay attention to, the American monetary policy and financial policy contradict each other. The monetary policy is aimed at raising the interest rate and eliminating the influence of the superabundance of money supply during the financial crisis. However, the unemployment rate is falling sharply in America, and they should take steps to prevent inflation. So Mr. Trump's fiscal policy is aimed at cutting taxes, which is a loose direction, but how does it affect the China's economy? There are few analyses in this report.

© Springer Nature Singapore Pte Ltd. 2018   55
Center for Macroeconomic Research at Xiamen University, *China's Macroeconomic Outlook*, Current Chinese Economic Report Series,
https://doi.org/10.1007/978-981-13-1005-8_5

Another question is the trend of China's economy. Most people have been predicting that China's economy will go down with a slowdown in growth rate over the past years, but it hasbeen increasing actually. What does it tell us if China's economy is falling down, and the growth is on a downward trend, but the IMF predicts that it will go up? The methodology tells us that if we want to forecast something, we need to look for its long-term trends and short-term fluctuations. As for China's economy, although it is likely to slow down in long term, but under the broad framework, short-term fluctuations may be more important. For instance, what will the fluctuations be in 2018 and 2019 in China? However, it is very difficult to predict this because it is random instead if fixed.

(2) The Basic Connotation of High Quality Development

Why do we need the high quality development? What is it? And how to achieve it? The high-quality development has two meanings: One is high quality, the other is development. Development is the most fundamental factor and also a necessary condition for the progress of human society. The high-quality development requires the respect for basic laws and principles. We are now entering the stage of high-quality development, which is an expression of the basic laws and principles. So, we must emphasize the respect for them during our specific work of implementing the high-quality development.

One example is the "Three Battles", which include preventing financial risks, alleviating poverty and protecting environment. What do the "Three Battles" have to do with China's development goals? For example, reducing financial risks now will possibly cause financial depression. Then what is the relationship between financial restraint and high-quality development? How can we minimize and eliminate possible financial repression in the process of guarding against financial risks? A typical statistics of financial repression is that our target of money growth and money supply growth was 12% at the beginning of 2017, but there was only an increase of 8.2% actually. How could it be able to comply with the big goal of development? It's two years left to reach the goal of fighting poverty by 2020, and our last way is using the financial support. Since a very important result that the goal that 100 million people get out of poverty had been achieving every year, we must reveal all the details in the last period. For environment protection, we already emphasized to avoid going the same way of treatment after pollution as western country, and the old way was the rule or law. The approach we now adopt is inspection by some team periodically, and then, what does it do with our development? In terms of high quality development, we should attach great importance to high quality, but development is more important.

Another example is the relationship between investment and consumption. According to the theories of Marx, Simple Reproduction and Expended Reproduction, it will be the simple reproduction if we only consider about consumption, and then how can we realize the expanded reproduction? Will we make progress without expanded reproduction? Only if consumption simulates investment, can it realize expanded reproduction. In other words, consumption leads to economic growth through investment, if not, it will be only the simple reproduction. Only the consumption which can simulate investment can form an expanded reproduction, as well as the economic growth.

High-quality development must be based on respecting the basic laws and the basic principles before we can better achieve the "Chinese dream."

5.2   Several Features of Economic Operation

Xianchun Xu
China Data Center, Tsinghua University, Beijing, China

National Bureau of Statistics, Beijing, China

China's economy kept stable in 2017, for example, the total output value had rebounded, the increase of consumer price had narrowed, corporate profits have increased, the economic structure had improved, and the new kinetic energy continues to accumulate. We can generalize the features of economic operation in 2017 through three perspectives: production, demand, and income.

(1) The Production

From the perspective of production, the economic operation in 2017 had three characteristics: Firstly, the value-added of tertiary industry played a major role in the development of economy; secondly, economic growth in 2017 was 0.2 percentage points higher than that of last year, and this was the first time that the economy growth rebounded since it had been declining from 2010; thirdly, although the GDP had a steady growth, that of some industries had an obvious fluctuation in the quarter.

As for the view of three industries: In 2017, the GDP increased by 6.9%, of which the primary industry was 3.9%, the secondary industry was 6.1%, and the tertiary industry was 8.0%. The tertiary industry with a contribution of 58.8% did the most to economic growth since it had a growth rate more than the GDP did. And among them, three industries, transport, storage and post, information transmission, and software information and information technology service industry, played a major role in the increase of it. For example, the industrial growth rate was 0.4 percentage points higher than that of last year, increasing to 6.4% from 6.0%. But that of construction was 2.9 percentage points lower, from 7.2 to 4.3% because of the decline in investment.

Among the four quarters, there were four industries whose added value fluctuates most. One was the construction, and its growth rate was between 3.1 and 5.4%. The second was finance, which was from 3.4 to 6.1%. The third was realty, of which the highest growth rate was 7.8% while the lowest was 3.9%. The last was software of information transmission and information technology service industry, and its growth rate had increased to 33.8% in the fourth quarter from 19.1% in the first quarter. Although the GDP had a steady growth, that of different industries fluctuated a lot.

(2) The Demand

From the perspective of demand, the growth of consumption demand slowed down in 2017 compared to 2016, as well as the contribution to economic growth. The same thing happened in investment demand. However, the import and export growth increased significantly than the last year, and the contribution of export to economic growth had turned positive.

Firstly, the growth of per capita cash consumption of China's residents in 2017 slowed down, both in nominal and real terms. And affected by general public budget outlays and public service expenditure, the nominal growth of government consumption increased compared to 2016, while its contribution to economic growth is 58.8%.

Secondly, the growth of investment demand in 2017 was 0.9 percentage points lower than that of 2016, and there was an obvious slowdown in new added capital growth. However, the inventory variation turned positive from the decline in 2016, and the finished goods' increase played the leading role, but it was still not enough to cover the decline of fixed-asset investment demand, since it was only a small proportion of inventory. The two investment demands jointly determined the situation of investment demand in 2017 and its contribution.

Lastly, the net export demand for goods and services turned positive from the negative in 2016, so did the contribution to the economic growth. According to the customs statistics of goods trade imbalance, the decrease improved in 2017 compared to 2016, mainly because the contrast between import and export price, since the import price increased by 3.9% and the export increased by 9.4%. The contribution of net export for goods and services was 9.1%, which was −9.6% in the same period of the last year.

(3) The Income

From the perspective of income, personal per capita disposable income maintained a steady and rapid growth in 2017, which was faster than last year, and the total profits of the industrial enterprises above designated size grew faster obviously.

Personal per capita disposable income represented a real increase of 7.3% in 2017, and it was 6.3%, 1 percentage points lower, in 2016.

The profits of the industrial enterprises above designated size grew faster obviously with an increase of 21%, 12 percentage points higher than that in the same period of the last year. There were three reasons: The industrial production sales growth rate accelerated. The production price increased obviously. And the unit costs decreased.

There were four reasons why the growth of general public budget rebound significantly in 2017: The economic growth, especially the nominal one picked up. The price increased, particularly in 2016, the PPI decreased. The corporate profits rose distinctly. And the price and amount of general imports rose together, so the total imports increased substantially, with more import linkage taxes, and causing the increase of growth rate of national general public budget receipts.

Above all, there were four most prominent features of China's economic operation in 2017: The first was that the drop in investment growth, especially the real investment growth, was quite apparent. The second was the PPI, which increased clearly from decline. The third was corporate profits, whether it was enterprises above designed size or service sector, they all had a pretty high growth rate. The last one was import and export of goods and services, which also increased from decline, and its contribution rate to GDP also turned positive.

5.3   The Connotation of China's Economic Development with High Quality

Yanbin Chen
School of Economics, Renmin University of China, Beijing, China

The party's nineteenth report had made a very important and new discussion about the trend of our country's economy: "China's economy had turned from the stage of high-speed growth to the stage of high-quality development". The Central Economic Working Conference convened at the end of December, 2017, had further pointed out that the essential features of the new era were "The transfer from the stage of high-speed growth to the stage of high-quality development". Promoting the high-quality development had great significance, and was not only the "necessary requirement for economy to keep the sustainable and healthy development", but also the "inevitable requirement to adapt to the changes of the social principle contradiction in our country and build a moderately prosperous society and socialist modernization country".

Theoretical community and all other communities interpreted the connotation of "High Quality" in many ways after the new concept was proposed by the Party Central Committee. In summary, the existing opinions thought that the connotations included relying more on innovation, higher production efficiency and economic benefit, better resources allocation and economic structure, putting more emphasis on the basic role of consumption in economic development, narrowing income gap, improving happiness as the guidance, caring more about guarding against the financial crisis, more environmental developing ways, and so on. The topics discussed were rich in detail.

Now the interpretation of "High-quality Development" is rich, but the surveys about relationship between them, especially the causality, are ongoing. The causality among variables is very important for economic researches. And if ignoring it, we may miss the nature of problems, even get the wrong conclusions.

The two typical examples, relationships between the doctors and infectious diseases and the cops and robbers, had explained a lot about the importance of causality. The characteristics and the descriptive indexes show that the more the doctors are, the more serious the infectious diseases are. And we may have a wrong conclusion that reducing the numbers of doctors can be helpful to treating the diseases if we don't understand the causality accurately. Actually, the disease is the cause while the doctor is the effect, and it is because of the serious diseases that we need more doctors in place. The cops and robbers are another interesting example. Hence, we need pay high attention to understanding the causality among variables when facing economic problems.

It will be difficult to find the breakthrough with less analysis of causality when surveying the connotation of "High-quality Development". For instance, better economic structure is the effect instead of cause of "High-quality Development". In its Seventh Five-Year Plan back in 1986, the Central Committee required to "adjust the direction and principle of industrial structure". The Fifteenth National Congress of the Communist Party of China held in 1997 said explicitly that "carrying out strategical adjustments of economic structure". And the Sixteenth, Seventeenth, Eighteenth and Nineteenth National Congress all affirmed the need to adjust and optimize the economic structure. The imbalance of economic structure is still a problem since it had been more than 30 years when "Adjustment of structure" was put forward for the first time, mainly because that the cause which obstructs the adjustment of structure had not been removed. And the connotations of "High-quality Development" which include relying more on innovation instead of elements like capital, better resources allocation, more environmental developing ways, etc., are the true cause. As long as we master these "causes", regarding as breakthrough, the industries with the serious pollution, considerable high consumption of resource and energy, will be dropped out naturally, and the industrial structure will also be adjusted naturally.

Also, putting more emphasis on the basic role of consumption in economic development is an effect rather than a cause. More than ten years ago, the government began to focus on expanding consumption. The report made at the Sixteenth National Congress of the Party pointed out: we should adjust the relationship between investment and consumption and increase the proportion of consumption in GDP gradually. The Seventeenth and Eighteenth reports also made the statement about speeding up the establishment of a long-term mechanism for increasing consumer demand. However, the consumption demand of residents had not been improved significantly, due to the same reasons of obstructing the growth of consumption which have not been removed, such as the gap between rich and poor, and the lack of financial expenditure for the people's livelihood. So, regarding the aim about narrowing the gap and increasing the financial expenditure for the people's livelihood as a breakthrough of "High-quality Development" can expand consumption effectively, thus enhancing the basic role of consumption in economic development.

Thus, it is not enough to discuss the connotation of "High-quality Development" only, we also need to focus more on the causality between different aspects, thereby grasping the impetus of "High-quality Development".

5.4 High Quality Development and Prevention of Systemic Risk

Kang Jia
China Academy of New Supply-side Economics, Beijing, China

Chinese Academy of Fiscal Sciences, Beijing, China

(1) High-quality Development

For high-quality development, there is already a consensus among the leaders, management department and academic, and even the practical work: The new development stage no longer emphasizes on the GDP. Then what should be emphasized on? And this a real question.

The government and management are in the right for political achievement. With the public power, they need an achievement of course, and should be doing something. But it is not enough to consider the GDP in the new stage since it had obvious fault and limitation, and the key is to form a reasonable performance evaluation index system with GDP and other necessary indicators. Although the Central Committee had said that China should not over emphasize on the GDP growth, the authoritative evaluating index had not been decided yet, and only if solving the problem as soon as possible, can it help to prevent the bias. Moreover, how to strengthen risk management? We need to understand that the essence of the Central Committee is guarding against system risk. The risk is everywhere actually, so everyone needs to work together to form a risk management evaluation system that is concise, usable and operational, and this is also a very practical problem.

(2) How to fight against systemic risk?

The substantive requirement of the Central Committee to guard against systemic risk, is to guard against the systemic and overall risk, and we can't use the one-size-fits-all approach to start fiercely defending, because it is a violation of the internal law. The one-size-fits-all approach can't guard against the systemic risk effectively since it exactly goes against the requirement of "treating differently and optimizing structure" which is the inherent request of supply-side structural reform. To prevent from systemic risk needs to be combine it with problem about how to deepen the supply-side structural reform and take the structural issues as the main part of contradiction, thinking more about how to find a reasonable, effective and sustainable mechanism of discrimination and paying more attention on solving the challenges of supply management through innovation. To optimize the structure of supply management, this is about the adaption to market discipline and necessary policy guidance of government. The government needs to protect property right and fair competition firstly, and then gives a directional guidance. Many times, we can take different steps before, during or after it, such as doing unpretentious things after and pretentious ones before. While following the market rules, it is also necessary to pursue a surprise attack and accomplish something with fewer mistakes. As for different fields, different industries and different companies, it won't be so simple to solve the problems only by the one-size-fits-all approach.

(3) **Effective Investment**

The potential of "effective investment" in our country can be still limitless, and we should lay stress on selective "smart investment" coupled with innovative mechanisms such as PPP. For example, Beijing had to develop the underground transport desperately, and carry out a lot of infrastructure construction, so does other cities with more than millions of people. The key is to link the funds well in large-scale and long-term investment, otherwise, it will be a mess. And if doing well, things will be easier with more legalized and standardized experiences accumulated. So, the key for PPP is that the Chinese government should push for higher level regulations as soon as possible, and legalized basis should be provided to the PPP project quickly. There must be problems of dealing with risks from the perspective of assets and liabilities, and the local government can't clear the debt entirely in this field, because when the government and the enterprises become partners, how could it be possible that all the debts are reflected in the enterprises? The standardized liabilities recognition of PPP won't prevent its progress, instead, it exactly helps to guide and inspire the healthy and sustainable development of PPP.

5.5   The Local Debt and Prevention of Financial Risk

Luolin Wang
Chinese Academy of Social Sciences, Beijing, China

The report is pretty thorough and realistic. For the prediction of economy in 2018, it said the growth of GDP slowed down slightly compared to 2017, I agree, but there are still some disagreements in economic academics. And I think some certain formulations in it are worth considering: For example, it said that the stability of the RMB exchange rate against the dollar would continue relieve the pressure of capital outflow in China, but I think the pressure will always be existing and very severe. Followings are some of my thoughts:

Firstly, what's the trend of risk development? Which one is the most important and pressing? I think the local debt risk is the most difficult problem to solve, due to the current investment system and administrative system in our country. Now some of measures adopted by central government for controlling the local debts may not mitigate the local debt risks. We predict that the risks will be controlled properly by 2018, but there are still many uncertain factors whether the risks will be bigger or smaller.

Secondly, how to control the financial risks? Our fiscal and monetary policies are basically in the form of "a tight, one loose, a tight, one loose and so on" in a long term, and the fiscal policy tends to be loose with tighter monetary policy. I'm very concerned about the potential risks existed in China's finance and I hope that all the financial experts here can consider this. Because now, spending in all aspects is increasing.

Thirdly, what is the status of China's economy in the international economy? Internationally speaking, it is uncertain. Judging from indicators like China's economic aggregate and infrastructure construction, China had been the world's second largest economy, but it is hard to judge which indicators and conditions can reflect the status of China's economy. About the status of China's economy, there are both positive and negative opinions in the world. We need to know that the opinion that China had been a developed economy is actually to make China be responsible for more than it should do.

5.6   Being Vigilant on the Continued Downward Risk of Economy

Liancheng Zhang
Capital University of Economics and Business, Beijing, China

The downward risk of economic growth in the short term is mainly from investment, consumption, export and the risk of a recession in the world economy once again.

The high growth doesn't mean low quality, and the low growth doesn't mean high quality either. The high-quality development at least contains three connotations: one is the green development. Another one is innovative development, also known as intensive development. The last one is letting people benefit from growth, otherwise, there will be a new imbalance between supply and demand.

(1)   The downward risk of economic growth in the short term

At the initial and medium term of industrialization, every country's economy might have high growth, but once it stepped into the last period of industrial process, the high growth was impossible to achieved. Overall the industrialization of China is still in its mid-stage, but the economy shows a continuous downward trend. And until the fourth quarter in 2017, the growth of economy had been dropping for 30 consecutive quarters since the second quarter in 2010, and the trend had not stopped. In the short term, the downward risk of economic growth comes from the continuous downward growth of investment, consumption and export, and maybe the shock of a recession in the world economy once again.

From the perspective of investment, China's investment, whether the national investment in fixed assets or the investment of private enterprises showed a cliff-type decline from 2009 to 2017, with no signs of stabilization: The growth rate of investment had decreased to 7.2% in 2017 from 30.5% in 2009, and we suppose that it may continue to decrease in 2018.

China also faces the pressure of capital outflow in the side of investment. From 2015 to 2016, only the capital account and the error and omission account on the balance of international payments were in deficit of 1.28 trillion dollars, and this was basically the whole scale of the actual use of foreign capital in China from 1979 to 2012. Since the severe capital control was imposed in 2017, things had improved, and there was a surplus of 15.6 billion dollars in the capital project, but the scale of capital outflow was still huge through underground channels. There still was a net outflow of 57.7 billion dollars in the error and omission account in the first quarter of last year, and pretty outflow funds won't appear on the balance of international payments.

From the perspective of consumption, the disposable income is a major term, because the consumption won't increase with no income growth behind it. Although the disposable income is increasing year by year, its growth rate continues to drop: It had decreased to 7.2% in 2017 since it was 12.2% in 2007, with no signs of a sustained recovery until now. As for the property incomes of the residents, one of them is the financial asset, whose income was shrinking drastically. The stock market made investors lost 32 trillion yuan only because the two-stock market crashed in 2015, and the GDP was only over 60 trillion yuan in 2015, which meant half of the GDP evaporated. The income of real estate also begins to shrink, and the trend had been going on which will always lead to a continuous decline in the growth rate of consumption, and it is very unlikely to drive sustained economic growth relying on consumption.

From the perspective of export, although the export growth becomes positive from negative this year, but the growth rate slowed down to −2.8% in 2015 from 31.3% in 2010. The growth rate was 10.8% in 2017, because the base of last year was low and the world's economy was at a high level. However, it is doubtful whether the export growth rate will continue.

From the perspective of external factors, the economies of America and the world are prone to crises every 8–10 years, may be 9 years on average. The last one was in 2009, and this year is 2018. There may be a new recession in American economy in 2019. And there have been some signs now: Three benchmark indexes in US had slowed down generally, which are the guide indexes 3–6 months ahead of the economic cycle. Until now, if the indexes continue to slow down rather than reaching a new peak, the recession will come soon. And the economic recession of America and the world might shock the China's economic growth once more.

(2) **Long-term economic growth trend**

For the long term of economic growth, we can know its trend through the potential economic growth rate. The potential economic growth rate of China is downward continually, so does the average growth rate, and the trend had not changed.

What determines the long-term economic growth of a country? The first is the number of inputs of production factors, including the capital inputs and nature resource inputs. It is the extensive growth of a country which is driven mainly by the capital and resource factors. The second is the efficiency of production elements, including technical progress, knowledge accumulation and human capital accumulation. The third is institutional factors containing system arrangement and system innovation, and they both have great influence on economic growth.

Why does the long-term trend of China's economic growth continue to slow down? For the factors inputs, the long-term reduction of investment growth rate must lead to the decline of growth rate of capital accumulation. Furthermore, the pressure of capital outflow, as well as the trend of diminishing marginal returns of capital, doesn't support the long-term sustainable economic growth, so does the reduction of labor supply, the decline of labor participation rate, the increase of labor cost, and the outflow and slow speed of human capital accumulation. And the TFP falls instead of rising, all of above show that the economic transformation still had a long way to go.

The key of economic transformation is the industrial transformation from extensive to intensive, rather than the transformation from industry to service industry. The failure of economic transformation comes from lack of microeconomic foundation of it and the necessary policy supports. Besides, the trends of "de-industrialization" and "counter urbanization" appeared in the past two years is also harmful to sustainable economic growth obviously. For example, Beijing government drives the low-end persons out of the city. Beijing isn't the only one, and many other prefecture-level cities follow the steps of it and do the same things. This is the "counter urbanization". It doesn't support the long-term sustainable economic growth of China whatever the purpose is.

From the perspective of institutional factors, the institutional bonus promoting the long-term rapid economic growth, are gone now, and the existing system had become a constraint on economic growth. So, we must promote the market-oriented economic system reform, including the enterprise system reform. However, in recent years, the government had taken a more dominating role in economic operations, and the degree of interference in the market and enterprises had become greater and greater. The trend of administrative monopoly, industrial monopoly and excluding competition is enhanced continuously, and the situation that the state-owned enterprises are becoming bigger increasingly to be a king is becoming increasingly acute. Why cannot we promote the market-oriented economic system reform?

5.7   Fighting a Hard Battle for Prevention and Control of Finance Risks

Bin Hu
Institute of Finance and Banking, Chinese Academy of Social Sciences, Beijing, China

(1)   How to understand the current financial risks?

Now everyone talks about financial risks, but there are still some misunderstandings and deviations in their understanding of risks. We must have a clear understand about that the financial risks that the Central Committee are considering to prevent and resolve aren't micro-level but systematic, macro-level and global. For example, the general secretary Xi Jinping had talked about the problems about financial risks and financial security in many occasions, he listed the major financial risks such as credit risk, shadow banking risk and other eight risks. Some of them may not necessarily require our great attention, for instance, the risk of a crime should not be the systemic risk we are concerned about. What's more, part of the debt risk of state-owned enterprises is still the micro risk. The question we should consider and see now is how to prevent the systemic risks resulting the damage of financial system.

(2)   The sources of systemic financial risk

What are the sources of systemic financial risk? One is periodic risk; the periodicity mainly means that the influence of economic cycle. The second is the institutional risk, which mainly is the risk of existing regulation system; the third is the structural risk, which is mainly the risk of imbalance in finance and the real economy. There are, in theory, three most core risks that China currently is fighting against: The first one is the systemic risk from the impact on financial system of periodic macroeconomic changes. Like it may cause the recession of real economy and then of the entire financial system. The second is what we called financial chaos, such as avoiding the regulation, innovating at random, establishing institutions casually and so on. The risks which include all the possible systemic risks in asset management, shadow banks and intersecting financial products, etc., are evolved for avoiding the regulation by the financial system and are away from the real economy. The third are the vertical, horizontal and external risks, which come from the regulatory of monetary policy and financial policy in the foreign countries, especially the developed countries, during the process of participating in globalization.

(3)   How to prevent and control financial risks?

The key to prevent and control the risks is the healthy development of economy itself. For the first source of financial risks, we need to consider how to adjust counter-cyclically and fight against the financial risks caused by real economy. One is the deleveraging of macroeconomy and real economy. The other is that of finance. Now the leverage ratio of residents is not too high, but that of financial department is very high, and this is the risk needed to be controlled. How can we prevent and control the risks in the process of deleveraging? The first is to reduce the enterprise burden, through not only the adjustment of tax burden, but also the consideration of how to decrease the cost and various charge of enterprise. Next is the financial regulation. How and where should we strengthen supervision? The next step may be related to the reform of national and organizational institutions, and the reform of financial regulation departments may be involved. How to build and choose an organizational structure of supervision, which is optimal and suitable in China and can solve the existing problem such as the regulation coordination issues, the supervision after the mixing and the coordination among departments, to prevent systemic risk? So, establishing mechanism of regulation coordination and making the existing Financial Development Committee practical are the core of strengthening supervision. Another is the behavior regulation. Our original regulation was the one based on compliance, then reached a micro-prudential one, and now we need the macro-prudential one. We offer to functional supervision and pay more attention on behavior regulation in order to prevent the vacuum of supervision, which means making the behavior and functional regulation practical. The People's Bank of China, along with the ten major ministries and commissions, publish new regulations, including new regulations for corporate management, which are all the reflects of the enhancement of existing functional and behavior supervision.

Besides, we need to focus on improving the flexibility and adaptability of regulation and making it more inclusive and forward-looking in the supervision. Why? Because to prevent the possible damage of financial system caused by over-regulation and strengthening supervision in the next step, we need a forward-looking preparation to keep the supervisor in a proper degree.

(4) **How to eliminating the financial risks?**

How to eliminating the financial risks? It needs the measures from reform. As it was highly emphasized at this meeting of the Central Financial Working Conference, we needed to eliminate the risks through reform and development, increase the degree of reform and opening up, and roll out more powerful measures for reform and opening up. How to solve the existing real economic problem with these measures being rolled out? As an example, how to stimulate the private capital, improve the contribution of it, or reduce its entry threshold to financial institutions? In order to make more private capital serve the real economy and control the financial risks at the same time, we need reform.

5.8   The Pressures of Downward Economic Growth and Institutional Reform

Ping Zhang
National Institution for Finance & Development, Beijing, China

Institute of Economics, Chinese Academy of Social Sciences, Beijing, China

(1)   The Pressures of Downward Economic Growth

Firstly, the prediction of Chinese economy may be worse than expected.

(1)   The comparative growth rate on moving base in the first half of 2017 exceeded the lowest growth rate of that in recent years, but in the second half, it was back to the lowest level in 6 consecutive years;

(2)   The pressure of economic growth was still relatively large, from the perspective of driving force of demand. The economic growth in 2017 mainly relied on the external demand which increased the economic growth by 1.09 percentages. The economic growth in 2017 was driven by the recovery of the international economy, but the domestic demand was still very weak;

(3)   The growth of consumption and income is slowing essentially, and the data of resident income growth published by the National Bureau of Statistics is strange. It is in doubt that the income growth of national resident was the same as that of rural residents, but more than that of the urban residents. The urban real consumption increased only by 4.1%. So, we should regard the domestic demand problem as a more serious one.

Most people think that the global growth will be better in 2018 compared to 2017. So, they upgrade their forecast for growth of this year, as well as the forecast for growth in China, for the reason that the recovery of world's economy will drive China's economic growth. For the goal of doubling China's GDP by 2020, current economic growth had already exceeded the lowest growth rate, 6.3%, required. Even so, I think we shouldn't ignore the pressures of downward economic growth.

Firstly, the pressure of downward economic growth in China is mainly due to the domestic demand, and its decline comes from the tightening of macroeconomic policy in the second half in 2017. (1) The first is the endogenous contraction of monetary policy: The old way of the central bank issuing money was dominated by foreign exchange, whose highest proportion was 83% in 2013 and 59% now. And the way to releasing money by the proportion of foreign exchange has already been out of date. (2) The second is the contraction resulting from regulatory factors. Various regulatory governments supervised in the spirit of "Father-loving" and aimed at letting the sons grow up easily. And now it is only for supervision, which we called the "Competitive Supervision". The CBRC pushed more than 1 hundred of regulatory letters within 6 months, and the CSRC, CIRC, financial ministry, EPA and the Land and Resources Bureau are all actively involved in the competition. The competitive supervision causes the shrink of all aspects such as the credit. (3) The last is the contraction caused by external uncertainty. The trade war between China and US and the appreciation of the exchange rate of RMB against the USD both discourage the exports of China, and so that Chinese economy can't maximize its benefits from the rebound of the global economy. The appreciation of the exchange rate has the contractionary effect and discourages the exports.

Secondly, the exchange rate of China needs to keep stable. The domestic capital cost increases actually along with following passively the step of America to increase the interest rate. So, if handled incorrectly, endogenous money supply will decrease as well as the competitive supervision, so the economic growth pressure is very large.

Thirdly, the decline of labor productivity growth and TFP increases the pressure of downward economic growth. And the decline presents structural characteristics: the proportion of industrial departments is decreasing, while that of finance and real estate are increasing, as well as the other service industries. And the proportion of construction industry increases dramatically, which is also an important reason for the decline of labor productivity growth. So, the most important thing to improve labor productivity growth is maintaining the market share of manufacturing industry, and the key is to improve their international competitiveness. The proper macroeconomic stability policies are necessary to stabilize the micro foundation. But we should consider the structural factors instead of using the one-size-fits-all approach.

(2) **The institutional reform**

The way out for China depends on reform. McKinsey measured the new economy of America that year and showed that the improvement of efficiency in American new economic stage came from not only the technical progress, but also the prosperity of competitiveness. The most important conclusion is that it is more significant to improve the economic efficiency by competition than by technical progress. We must enforce the market reform and property protection.

For local debts, the main problem is that the financial system reform is not in place, with 83% of the authority but only 50% of financial power, and how to make up the shortfall in funding? The public services every taxpayer enjoys should match their paying step by step with all the tax burdens being transferred to enterprises and all the gaps between public expenditures and public incomes being left to local government. If we don't go further to reform the financial system, it will be hard to rectify the behavior of local government as well as state-owned enterprises. So, the biggest problems of reform in China, including financial reform, are all related to the transition and reform of financial system.

5.9 Innovation Actuation and the New Kinetic Energy in China's Economy

Yansheng Zhang
Academic Committee, National Development and Reform Committee, Beijing, China

China Center for International Economic Exchange, Beijing, China

The root of the replacement of old growth drivers of China's economy is innovation. And in order to speed up the transformation of new and old system, we need to promote the structural transition in more places and the new technologies, new industries, new formats and new models in all localities and departments.

(1) The Innovation Actuation

There is development imbalance of innovation actuation between the Three regions of China now.

The first region is the eastern region. The R&D intensities in the Yangtze River Delta, the Pearl River Delta and the Guangdong, Hongkong and Macau are all above the average level of OECD. The R&D inputs of Guangdong is 2035 billion yuan in 2016, while Jiangsu is 2026 billion yuan, as well as Beijing and Shanghai, their inputs and intensities of R&D both show that they have entered the innovation stage. With ten or twenty years, R&D stocks will accumulate to a certain extent, and this region will surely have world-class universities, which will present a new trend driven by new industries, new technologies, and new businesses.

The second region is the central regions with 14 provinces in. The R&D intensities of Hubei is 1.86 while Sichuan is 1.72. The economies of these 14 provinces in this region are still in the different stages of investment-driven development.

The third is the western regions with 11 provinces, which are relatively backward, the R&D intensities in this region are all relatively low, and all of their R&D inputs are only 880 billion yuan.

(2) The major question of transformation the kinetic energy of China's economy

The first question is the unbalanced and insufficient problems. In the three regions, there are three different economic development stages: innovation-driven, investment-driven and resource-driven. How to address the gap between them by policy and strategy in terms of the reform regarding quality, efficiency and power.

The second question is the R&D of the manufacturing industry. The R&D intensity of China's manufacturing industry is still much lower than that of Europe, America and Japan. The R&D intensity is lower than 0.5 for most traditional manufacturing, even that of the equipment manufacturing and heavy chemical industry is only between 1 and 2.

The third question is that the proportion of basic research in R&D innovation in China is 12 percentage points lower than that of America. Whether the universities or the scientific research institutes are in the industrial economy and have not entered the stage of knowledge creation. So, while the report made at the Nineteenth National Congress of the Party also said the high-quality development, how can we finish the transformation of kinetic energy mechanism during the ten-years transition?

The forth question is that whether the private investment or the foreign investment is in a bottleneck. As a result, the environmental reform like business environments, investment environments, market environments, policy environments, innovation environments and so on, is imminent. The Nineteenth report suggested to build the free-trade experimental areas and pilot free trade ports in several places. Compared to Hongkong and Singapore, the business environment in China ranks at 78, Japan ranks at 32, and Taiwan ranks at 32 while Hongkong ranks at 5, and Singapore is the NO. 1 or 2. The ranking of China's business environment doesn't match our position as the second largest economy. And there is no rise in the ranking in 2017 and 2018, while the ranking of India increases more than by 30 in 2018 compared to its ranking of 130 in 2017. Above all, we need more competition in both business environment and investment environment. For the ownership structure, the proportion of private enterprises in China is 61.8%, so how can we form a mixed ownership system in service and other key fields?

The last one is the competitive supervision. A core problem is how can we make progress in modernization of governance system and capacity. Chinese government should act and supervise in a modern way. And Shenzhen had tried to do like this once, but the government was not willing to do it, which was the biggest obstacle. It is a test again since Shanghai also wants to try now. In order to achieve modernization of governance capacity and provide better public services and products, we must do it in the way of modern governance. China is on the node of the transformation of the new and old systems. If the institutional model is unchanged, then the policy environment is also hard to change.

5.10   New Foundations Formed Preliminarily, and New Features Appearing Gradually—The Analysis and Forecast of Economic Situation in 2018

Liqun Zhang
Macroeconomic Research Department, Development Research Center of the State Council, Beijing, China

China's economic growth had been slowing down from 2010 to 2016 and decreased to 6.7% from 10.6%. The main reason of policy was that the "package plan" exits completely. And the main reason for market was the international financial crisis, which caused the export growth continues to decrease significantly (from 2010 to 2016, the export growth rate in dollars decreased to −7.74% from 31.3%). And the investment growth rate continued to decrease significantly as a result of imbalance of domestic urbanization promotion (from 2010 to 2016, the investment growth rate of total social fixed assets decreased to 8.1% from 24.5%).

(1) **New foundations formed preliminarily**

Firstly, the export situation of our country begins to improve with the external economic environment stabilizing. In 2017, the international trade growth rate began to be higher than growth rate of global GDP, and the export situations of developed countries and new developing regions all improved, all of which could be an important signal of the economy getting rid of the downturn in growth. Although there are still many uncertain factors, the foundation to judge whether the Chinese export growth rate, which decreased in the past, hasbeen steady, by discussing the recovery trend of world's economy and the effect of transformation and upgrading of China's export enterprises has already been formed preliminarily.

Secondly, the imbalance of domestic urbanization promotion is being solved now, and the positive factors supporting the investment growth also increase. The decline of investment growth rate was initially due to the exit of "package plan" and that infrastructural investment growth rate continued to decrease significantly. After the growth stabilization policies in 2012, the infrastructural investment growth rate increased significantly, and its effect to stabilize the investment growth was also enhanced. Then the investment growth rate continued to decrease, mainly because of the significant decline of investment growth rate of real estate, which mainly came from the two difficulties: difficult to get the land and difficult to sell the house. And it had a close relationship with the imbalance of urbanization promotion. The urban popularity improved steadily and the number of people increased steadily in the process of urbanization. Since the development of basic conditions, such as infrastructure, public services and market environment, is unbalanced in different cities, the improvement of production city integration level is unbalanced, and then there is a great difference in rate of improvement of urban popularity, as well as the imbalance of urban population growth. There are also two extreme cases that the big cities are full of people while few people are in the small and medium-sized cities. The more rapidly the popularity grow, the higher the growth of real estate will be, and it must attract enterprises of real estate. After a certain period of time, there will be some more and more serious problems called "difficult to get land" in the big cities with a fast-rising popularity. On the other hand, because of the "constructing cities" appeared after 2010, there are inclination of only enlarging the area of built-up area without considering the systemic conditions about the improvement of urban popularity. The local government pushes a lot of preferential policies, especially the policies of land supply, to attract the attention of real state enterprises, especially those which can't get the land in big cities, and many of them get land and build houses in small and medium-sized cities. Because the development doesn't match the conditions to support the improvement of urban popularity, there are big imbalances between housing construction layout and the urban population growth pattern, and then the serious problem called "difficult to sell the house". Under the constraint that there are pretty serious difficulties in the circulation of funds, but the regions with big market potential can't get the land, the investment growth rate of real estate continues to decrease significantly (from 2010 to 2015, it decreased to 1% from 33.2%), which restricted the marketing of commodities like heavy chemical raw materials and energy, and the related enterprises' operating rate decrease and their benefits continue to drop. So, it is the main reason why the investment growth rate decreases that the decline of investment growth rate of real estate caused by imbalance of urbanization. The population in a number of second-tier cities grows rapidly in recent years. The first-tier cities begin to develop towards the urban agglomeration with the surrounding small and medium-sized cities. And driven by the central cities, the popularity of some third-tier or forth-tier cities is also improving. According to the strategic deployment of new urbanization published by the Central Committee, the work of long-term and comprehensive planning of the city is being strengthened. The underground infrastructures speed up to be constructed. The internal infrastructure of urban agglomeration and urban integration of public services are being pushed actively. The problems of imbalance of urbanization promotion and inadequate development is being solved now. As a result, the problems of "difficult to sell the house" in the first tier cities, hot second-tier cities and a number of third-tier or fourth-tier cities are relieved, and so do the problems of "difficult to get land" in those second, third or fourth-tiers cities. All of above encourage the recovery of development and construction activities, as well as the improvement of investment growth rate of real estate. The recovery and activity of real estate and infrastructure have driven the better situation of the marketing of commodities like heavy chemical raw materials and energy, and higher price and profit of them, as well as the rebounds of manufacturing and private investment. The foundation that the investment growth rate, which decreased in the past, has become steady, is formed preliminarily.

Thirdly, economic transformation and upgrading have made great progress, and the reform achieves positive results. Depending on the judgement that major stage changes of the development environment has taken place, the Party and government have adjusted the development ideas and the focus of policy since 2012. The macroeconomic adjustment changes the traditional thoughts of imposing policy counter-cyclically and makes us pay more attention on adjusting the structure, deepening the reform and try hard to increase the ability to adapt the development environmental changes. Especially it sets the systemic task to promote the structural reform of the supply side, and clearly deploys the five tasks of "cutting overcapacity, reducing excess inventory, deleveraging, lowering costs, and improving areas of weakness". In recent years, the market and the government work together to speed up cutting excessive capacity. The process of reducing inventory achieved outstanding results that the commercial housing area for sale decreased 15.3% year on year at the end of December 2017, with a commitment to treating the new urbanization as our main task and taking multiple measures at the same time. There were also significant results in the process of deleveraging and lowering the cost, and the debt-to-asset ratio of industrial enterprises above designated size was 55.8%

at the end of November, 0.5 percentage points lower than the same period of last year. The cost of enterprises keeps dropping, and the cost per 100 yuan of main business income of industrial enterprises above designated size was 85.26 yuan from January to November, 0.28 yuan lower than the same period of last year. The investment in areas of weakness speeded up. The ecological protection and environmental governance investment rose by 23.9%, 16.7 percentage points faster than the total investment. The water management industry investment increased by 16.4%, 9.2 percentage points faster than the total investment, so did the agriculture investment. Under the pressure of market competition, the scientific and technological R&D activities of industrial enterprises increased significantly, research and development expenditure rose to 2.05 percent of GDP from 1.79%. High-tech manufacturing investment increased by 17%, an increase of 2.8 percentage points over the previous year. Added value of high-tech industry rose by 13.4%, 6.8 percentage points faster than the industry. The enterprises became more and more accustomed to the slowdown growth by controlling the cost, improving quality of products and services, and increasing the concentration through business acquisitions and reorganizations. Overall, the supply-side structural reform makes the excessive-supply problem is eased, and enterprises adapt to the new development environment increasingly.

Together, these factors constitute a new foundation for economic growth and support the steady growth of economic growth. This also marks the end of the six-year-old pattern of economic growth.

(2) New Features Appearing Gradually

The demand of domestic and foreign markets grows steadily in 2018. The potential of supply growth is great while market-oriented and legalized activities of cutting production overcapacity tends to be standardized, and the production capacity of main products is ample. Then we predict that there will be a balance of total market supply and demand in 2018, with no big gaps. And affected by the features of market supply and demand, the expected economic growth will be steady in 2018, and the price will increase steadily too. Also we predict that the growth rate of GDP will be roughly the same as 2017. The PPI will decrease significantly compared to 2017, while the increase of CPI will be 2%. And this will be a good macroeconomic environment for transferring into high-quality development. With the supply-side structural reform being pushed further, the spirit of the Nineteenth Congress is being fully implemented, and the new features of economic growth, with improving quality and increasing efficiency, reducing emission and saving energy as the main task, will be evident increasingly.

Chapter 6
Questionnaire Survey
on the Macroeconomy of China in 2018

To keep abreast of the macroeconomic situation and policy trend, an annual questionnaire survey of China's macroeconomic situation and policy jointly started twice a year since the first time in August 2013, held by *the Economic Information Daily*, Xinhua News Agency and the Center for Macroeconomic Research at Xiamen University (one of the Key Research Institutes of Humanities and Social Sciences of the Ministry of Education of China). This is the tenth-time questionnaire survey about the study. There were 22 questions directly about China's macroeconomic situation and policy trend in the questionnaire, and we invited some domestic economists for this survey in January 2018, and finally got responses from 128 of them. This survey offered the latest understandings and judgments of experts concerning the economic situation of the world, trends of some major indicators about China's macroeconomy, and trends of China's macroeconomic policies in 2018. The results of this survey are reported as follows:

6.1 The Economic Situation of the World in 2018

In accordance with the latest economic forecast of the IMF on January 22, 2018, the economic growth rate of the US would be 2.7% in 2018, and 0.4 percentage points higher than the forecast in October, 2017. Hence, we conducted a questionnaire survey on the trend of USA's economic growth in 2018. The survey reflected that 43% of the experts answered that the economic growth rate of USA would be "between 2.5 and 2.7%" in 2018. 43% claimed that it would be "between 2.7 and 2.9%". 10% claimed that it would be "between 2.9 and 3.1%". 3% claimed that it would be "2.5% or less". 1% of the experts thought that it would be "3.1% or

© Springer Nature Singapore Pte Ltd. 2018
Center for Macroeconomic Research at Xiamen University, *China's Macroeconomic Outlook*, Current Chinese Economic Report Series,
https://doi.org/10.1007/978-981-13-1005-8_6

more". Overall, more than half of the experts believed that the US economy in 2018 would show a warming trend, and the situation is optimistic.

The IMF revised up the forecast for the economic growth rate of the Eurozone on January 22, 2018, to 2.2% in 2018, compared with the forecast in October, 2017 as 1.9%. We also conducted a questionnaire survey on the trend of Eurozone's economic growth. The survey reflected that 52% of the experts estimated the number would be "between 2.0 and 2.2%" in 2018. 44% claimed that it would be "between 2.2 and 2.4%". 3% claimed that it would be "2.0% or less". And 1% claimed that it would be "between 2.4 and 2.6%". In sum, more than half of the experts believe that the economy growth rate of Eurozone would show a downward trend in 2018, but more than 40% of the experts expressing a relatively optimistic expectation believed that the economy growth rate of Eurozone would show an upward trend and economy would be in recovery in 2018, and the situation is optimistic.

Since 2017, international commodity prices ended higher after initially falling. In the first half of 2017, since the impact of international oil price, international commodity prices showed phased shock downward, thereafter it was rally, and annual general level in 2017 had a more modest rise compared to 2016. How about the trend of commodity prices in 2018? The survey showed that 81% of experts expected the commodity prices would show a modest upward trend, 11% considered that it would be oscillation, 6% believed that it would be a strong rebound, and 2% thought that it would show a downward trend. In sum, nearly 90% of the experts believed that the international commodity prices would show an upward trend in 2018.

In 2017, with the gradual recovery of the European economy, the Eurozone began to get rid of deflation. The euro continues to maintain the trend of appreciate against the dollar, and the exchange rate of USD/EUR is 1.2262 on January 22, 2018. Hence, we conducted a questionnaire survey on the variation trend and range of the exchange rate of USD/EUR at the end of 2018. The survey reflected that 62% of the experts estimated that "it would keep stable". 31% of the experts claimed that the euro would continue to appreciate against the dollar, and at the end of 2018, the exchange rate of "USD/EUR would be between 1.2 and 1.4", of which the mean is about 1.2675. Another about 7% of the experts believed that the euro against the dollar would turn into depreciation, and the exchange rate of USD/EUR by the end of 2018 would be between 0.9267 and 1.18, of which the mean is about 1.0823. Overall, more than 60% of the experts believed that "the exchange rate would remain stable", and there are about 30% of the experts claimed that "the euro would continue to appreciate against the dollar" in 2018, but the change would not be large.

6.2   The Forecast of Some Major Indicators of China's Macro-economy in 2018

According to the preliminary data released by China National Bureau of Statistics on January 18, 2018, growth rate of China's GDP in 2017 was 6.9%. How about the growth rate of China's GDP in 2018? The survey showed that 61% of the experts thought that it would be "between 6.7 and 6.9%", 25% expected that it would be "between 6.4 and 6.6%", 12% chose "between 7.0 and 7.2%", and each of 1% claimed that it would be "6.3% or less" and "7.3% or more". In sum, because the GDP increased by 6.9% in 2017, more than 90% of the experts considered China's economic growth would fall in 2018.

In 2017, China's CPI increased by 1.6% compared to 2016, and was 0.4 percentage points lower than that in 2016. How about the trend of China's CPI in 2018? The survey showed that 76% of the experts expected the growth rate of China's CPI would be "between 1.6 and 2.0%", 11% thought that it might be "between 1.1 and 1.5%", 8% held the view that it would be "between 2.1 and 2.5%", 4 and 1% claimed that it would be "2.6% or more" and "1.0% or less" respectively. In sum, because the CPI increased by 1.6% in 2017, nearly 90% of the experts forecasted a certain upward trend of CPI in 2018.

In 2017, China's PPI increased 6.3% compared to the same period of 2016, it ended a five-year downward trend since 2012. How about the trend of China's PPI in 2018? The survey showed that 43% of the experts expected that the growth rate of China's PPI would be "between 6.3 and 6.7%", 28% of the experts expected it would be "between 5.8 and 6.2%", 23% of the experts claimed that it would be "5.7% or less", 5% of the experts believed that the growth rate of PPI in 2018 would be "between 6.8 and 7.2%", and 1% of the experts thought that it would be "7.3% or more". In sum, since the PPI increased by 6.3% in 2017, nearly half of the experts believed that the growth rate of China's PPI in 2018 would show an upward trend, but more than half of the experts thought that it would show a downward trend.

The exchange rate of the CNY(RMB) to USD (CNY/USD) was 6.5342 by December in 2017, which maintained the trend of strong. About the exchange rate of the CNY(RMB) to USD in 2018, 61% of the experts expected the CNY against the dollar would remain stable. 27% of the experts expected that the CNY(RMB) against the dollar would continue to appreciate, and at the end of 2018, the exchange rate of the CNY(RMB) to USD (CNY/USD) would be between 6.15 and 6.54 Yuan, 6.31 Yuan on average (this question is disparity filling). 12% of the experts believed that the CNY against the dollar would begin to depreciate, and at the end of 2017, the exchange rate of the CNY to USD would be between 6.58 and 6.8, 6.6268 Yuan on average (this question is disparity filling). By December, 2017, the exchange rate of the CNY to USD was 6.5342. Therefore, nearly of 90% the experts believed that the exchange rate of the CNY to USD would be appreciate in 2018.

In 2017, China's investment in the fixed assets (excluding farmers' investments) was about 63.1684 trillion RMB, a nominal increase of 7.2%, 0.9 percentage points lower than the same period of 2016 and kept the same as January to November. How about the nominal growth rate of China's investment in the fixed assets in 2018? The survey showed that 46% of the experts expected the total fixed investments would increase year-on-year at the rate "between 6.1 and 7.1%", 45% considered that it would increase at a rate "between 7.2 and 8.0%", 6% chose "between 8.1 and 9.0%", 2% thought that the growth rate would be "6.0% or less" and 1% thought that the growth rate would be "more than 9.0%". In sum, since the nominal growth rate of China's investment in the fixed assets in 2017 was 7.2%, nearly 50% of the experts believed that the growth of China's fixed investments would continue to decline in 2018, and more than 50% of the experts believed that the growth of China's fixed investments would begin to rise in 2018.

In the 2017, the private investment in fixed assets was 38.1510 trillion RMB, a nominal increase of 6.0%. Private investment accounted for 60.4% of the total and was 0.1 percentage points lower than that from January to November. Private investment growth continued to be lower than the national fixed asset investment growth, leading that the proportion of private investment for total investment continued decline. Thus, we investigated the reason (this question is multiple choice). The survey showed that 72% of the experts agreed that the lack of effective investment targets, the downturn of macro environment, plus the shrink and the falling profits of the traditional manufacturing industry, were leading to few effective and profitable investment targets for private capital. 72% of the experts agreed that the new type of financing was difficult and expensive, and many policy-related factors led more credit to flow into state-owned enterprises or return to the financial system, which resulted in the "crowding out effect" on private capital, together with the debt default problems of many private enterprises making financing more difficult and expensive. 64% of the experts claimed that private enterprises lacked confidence in the economy and the market outlook. 56% of the experts agreed that many industries such as the tertiary industry were still under control, with the investment areas being limited, the effective investment channels being narrow or pool, and the approval procedures for investment being cumbersome. 42% claimed that the sustained upward trend of house prices squeezed the development space of the real economy and the private investment. Other reasons from the survey included that the sustained upward trend of house prices squeezed the development space of the real economy and the private investment excessive liberalization in the financial sector, poor supervision and higher financing costs, property rights protection and business environment are not perfect, lack of effective protection of private enterprise property rights and the legitimate rights and interests of private entrepreneurs and so on.

In 2017, China's investment in the real estate sector was about 10.9799 trillion RMB, a nominal increase of 7.0% over 2016, and the growth rate was 0.1 percentage points faster than that of last year and 0.5 percentage points slower than that of January to November in 2017. Among them, residential investment was 7.5148 trillion RMB, an increase of 9.4%, and the growth rate was 0.3 percentage slower than that of last year. Residential investment accounted for 68.4% of the total investment in the real estate. How about the growth rate of China's investment in the real estate market in 2018? The survey showed that 62% of the experts expected it would be "6.1 and 7.0%", 19% expected that it would be "7.1 and 8.0%", 13% claimed that it would be "6.0% or less", 5% claimed that it would be "8.1 and 9.0%", 1% expected that it would be "more than 9.1%". In sum, compared to in 2017, there were 75% of the experts believed that China's real estate market in 2018 would show a slowdown in investment, and 25% of the experts believed that China's real estate market investment would keep it steady or show a warming trend in 2018.

According to 2018 national economic operation report released by the National Bureau of Statistics on January 18, 2018, five priority tasks were promoted solidly and policy results continued to show. In 2017, the national industrial capacity utilization rate was 77.0%, and it was highest in the past five years. So, How about the current situation in China's industrial sector? The survey showed that 40% of the experts agreed that "The speed of de production is moderate and in good condition", 28% of the experts agreed that "The speed of de production is little fast, should be adjusted", 20% of the experts agreed that "The speed of de production is slow, and affect the structure reform on supply side", 9% of the experts thought that "The speed of de production is a fast, and should be careful", and 3% of the experts thought that "The speed of de production is too slow, and overcapacity problem is serious". In sum, nearly 40% of the expects thought that the speed of de production in industrial sector was moderate, leading a great result, and about 30% of the experts claimed to slow more, but more than 20% claimed to speed.

In 2017, the annual per capital disposable income of the whole country had a nominal increase of 9% compared to 2016, it had a real increase of 7.3% after adjusting for inflation, compared with the real increase in 2016 as 6.3%. According to the permanent residence, per capita disposable income for urban residents increased by 6.5% and that for rural residents increased by 7.3%, and the income for rural residents increased faster than that for urban residents. Hence, we conducted a questionnaire survey on the growth trend of the income of the rural and urban residents at the end of 2018. The survey reflected that 39% of the experts claimed the per capita disposable income for the urban and rural residents would increase by "6.5% or more" and by "7.3% or more" respectively, 32% of experts claimed that the per capita disposable income for the urban and rural residents would increase by "6.5% or less" and by "7.3% or less" respectively, 20% of the experts claimed the per capita disposable income for the urban and rural residents would increase by "6.5% or less" and by "7.3% or more" respectively, 9% of the experts claimed the per capita disposable income for the urban and rural residents would increase by "6.5% or more" and by "7.3% or less" respectively.

In 2017, China's total social retail sales of consumer goods were about 36.6262 trillion RMB, a nominal increase of 10.2%, and the growth rate was 0.2 percentage points slower than that of last year. How about the growth rate of China's total retail sales of consumer goods in 2018? The survey showed that 53% of the experts expected the total retail sales of consumer goods would increase year on year at the rate "between 9.7 and 10.2%", 38% considered that it would increase at a rate "between 10.3 and 10.8%", both 4% chose "9.6% or less" and "between 10.9 and 11.4%", and 1% of experts held the view that it would be "more than 11.5%". So, the results reflected that compared to 2017, about 60% of the experts thought the growth of the total retail sales of consumer goods would show a decline trend in 2018. However, more than 40% of the experts maintained that it would increase in 2018, and the role of consumption on economic growth would gradually be stimulating.

In 2017, China's total exports were about 15.3318 trillion RMB, a nominal increase of 10.8%. How about the growth rate of China's total exports in 2018? The survey showed that 38% of the experts expected the total exports in dollar terms would increase "between 10.3 and 10.7%", 32% considered that it would increase at a rate "between 10.8 and 11.2", 22% chose "10.2% or less", 5% held the view that it would be "between 11.3 and 11.7%", and 3% chose "more than 11.8%". In sum, the results reflected that 60% of the experts believed that the growth of the total exports would show a degree of decrease in 2018, and 40% of the experts believed that the growth of the total exports would show a degree of rise in 2018.

In 2017, China's total imports were about 12.4603 trillion RMB, a nominal increase of 18.7%. How about the growth rate of China's total imports in 2017? The survey showed that 39% of the experts expected the total imports in dollar terms would increase "between 18.7 and 19.1%", 29% considered that it would increase at a rate "between 18.2 and 18.6%", 22% chose "18.1% or less", 9% held the view that it would be "between 19.2 and 19.6%", and 1% chose "more than 19.7%". In sum, the results reflected that 50% of the experts believed that the growth of the total imports would show a degree of rise in 2018, and 50% of the experts believed that the growth of the total imports would show a degree of decrease in 2018.

6.3   The Macroeconomic Policies May Be Taken by China in the Future

By December in 2017, the balance of China's broad money supply (M2) was 167.68 trillion RMB, an increase of 8.2% over the same time in last year, 0.9 and 3.1 percentage points lower than that of last month and last year respectively. How about the growth rate of China's broad money supply (M2) in 2018? The survey showed that 43% of the experts had the expectation that it would grow "between 7.6 and 8.1%". 38% considered "between 8.2 and 8.7%". 13% considered "between 8.8 and 9.3%". Each 3% of experts held the view that it would be "7.5% or less" and "more than 9.4%". In sum, the results reflected that more than 50% of experts thought that the growth rate of M2 would be "more than 8.2%" in 2018. It probably means the government of China would keep a moderate loose monetary policy in 2018.

According to the preliminary statistics of the People's Bank of China, in 2017, the scale of new yuan loans was up to 13.53 trillion RMB in China, 878.2 billion

RMB higher than the same period of 2016. Then, how will experts comment on where those new yuan loans go (this question is multiple choice)? The survey showed that 72% of the experts believed that new yuan loans didn't change the difficult and expensive situation of financing for private investment, 64% of the experts believed they ensured the expansion of investment in the field of infrastructure, 56% of the experts believed they accelerated investment growth of state-owned and state-controlled enterprises, 52% of the experts believed they didn't enter the real economy, leading to continuous drop in investment growth, 40% of the experts believed credit expansion was difficult to play effect due to poor monetary policy transmission channel, and 36% believed they stimulated the expansion of the real estate market.

According to the Central Economic Working Conference, China would follow a pro-active fiscal policy, adjust and optimize the structure of fiscal expenditure, and enhance fiscal sustainability. China's economy had shifted from high speed to high quality development, how about the implement of China's fiscal policy in 2018? Hence, we conducted a questionnaire survey on the implement of China's fiscal policy in 2018 (this question is multiple-choice question). The survey showed that 80% experts claimed whether it was the tax policy or the structure of the fiscal expenditure, the fiscal policy should pay more attention to the high-tech industries, the precision poverty alleviation and the rural revitalization. 65% of the experts thought that it should implement structural tax preferential policy, such as further reducing the preferential tax rate and expanding identification scope of new and high technology enterprises. 57% of the experts held the views that it should increase the openness of the infrastructure field, encourage local governments to provide more preferential policies for the participation of private capital in the PPP project combined local realities, and let PPP project return to its source. 54% of the experts thought that it should further clean up non-tax revenue, partly cancel government-managed funds and lower some standards levied by the government. 50% of the experts believed that it should continue to implement and improve the policy of "camp to add" pilot and enlarge the effect of tax reduction. 42% the experts claimed that it should tighten the general budget and loosen the generalized finance simultaneously and increase the newly increased scale of special bond by a big margin, and the budget deficit rate will not rise further. 22% of the experts believed that the added-value tax should give first place to improve and expand the scope of deduction.

The Central Economic Working Conference was held in Beijing from December 18 to 20, the conference was focused on the economic work of 2018, and we investigated the most prominent issues (this question is multiple choice). The survey showed that 87% of the experts held the view that promoting economic growth from high-speed growth to high-quality development was the fundamental requirement for determining development ideas, formulating economic policies, and implementing macro-control in the present and future periods; 60% of the experts believed that we should accelerate the establishment of a housing system reform and a long-term mechanism which was multi-subject supply, multi-channel guarantee, and rent-purchase combined; 54% of the experts believed that rural revitalization strategy would become the focus of rural work; 47% of the experts believed that seeking improvement in stability was still the main tone and needed long-term adherence; 40% of the experts believed that we should promote the formation of a new opening-up pattern and 20% of the experts believed that we should promote state-owned capital to become stronger and bigger.

The 19th CPC National Congress and Central Economic Working Conference both pointed out that China's economy had shifted from high-speed growth to high-quality development, thus we investigated on the basic connotation of promoting high-quality development of China's economy (this question is multiple choice). The survey showed that 76% of the experts held the view that we should focus on accelerating the construction of an industrial system that combined the development of the real economy, technological innovation, modern finance, and human resources; 75% of the experts believed that we should focus on solving the imbalance of supply and demand in the real economy, the imbalance between the financial sector and the real economy, and the imbalance between real estate and the real economy; 71% of the experts believed that we should improve the quality of the supply system and promote the continuous improvement of the quality of goods and services; 57% of the experts believed that we should let green development become a universal form, and form a new pattern of modernization construction in the harmony between man and nature; 53% of the experts believed that we should strengthen the protection of intellectual property and make innovation the first driving force for development; 50% of the experts believed that we should improve the socialistic market economy system, improve the property right system, give better play to the role of the government, continue to deepen reform and opening up, and promote the formation of a new pattern of comprehensive openness and 42% of the experts believed that we should implement a rural revitalization strategy and a regional coordinated development strategy.

In addition, we also investigated on the ultimate risk to the Chinese economy (this question is multiple choice). The survey showed that 75% of the experts held the view that the problem of state-owned enterprises and local government bonds was serious. As a result, the level of bad debts in the banking industry would increase, and it would lead to systematic financial risk; 51% of the experts believed that real estate price bubble was serious and there was a large financial risk; 48% of the experts believed that the economy lacked new growth points, the economic growth mechanism was slow to change, innovation had not yet become the primary motive force for economic growth, and the space for new industry development was limited; 44% of the experts believed that the growth rate of private investment continued to be lower than the growth rate of state-owned investment, and the increase in private investment growth was lacking in sustainability; 30% of the experts believed that the over-expansion of shadow banking system had led to high leverage of state-owned enterprises and lack of effective supervision measures. In addition, a few experts have proposed other risks that China's economy is currently facing, including that total TFP growth slows, income distribution gap further widen, international financial risk causes exodus of Chinese capital, excessive income gap leads to insufficient economic growth and so on.

Appendix
The Forecast Made by the 128 Experts on Major Indicators of China's Macro-economy

Major indicators of China's macro-economy in 2018	Forecast by research team (%)	The interval and ratio of forecast by experts (%)	
		Interval	Ratio
The growth rate of China's GDP	6.73	6.7–6.9	61
		6.0–6.6	25
The growth rate of China's CPI	2.13	1.6–2.0	76
		1.1–1.5	11
		2.1–2.5	8
The growth rate of China's PPI	4.64	6.3–6.7	43
		5.8–6.2	28
		5.7 or less	23
Nominal growth of China's total retail sales of consumer goods	10.60	9.7–10.2	53
		10.3–10.8	38
Nominal growth of China's fixed investments	6.57	6.1–7.1	46
		7.2–8.0	45
Nominal growth of China's total exports	9.65	10.3–10.7	38
		10.8–11.2	32
		10.2 or less	22

© Springer Nature Singapore Pte Ltd. 2018
Center for Macroeconomic Research at Xiamen University, *China's Macroeconomic Outlook*, Current Chinese Economic Report Series,
https://doi.org/10.1007/978-981-13-1005-8